D 553.453 SLA

8000065577
TREMOUGH CAMPUS LIBRARY

CAMBORNE SCHOOL
ES

AF841557

PLUS LIBRARY
WITHDRAWN

Mineral Resources
Consultative Committee

**Mineral Dossier No 9**

# TIN

Compiled by
D Slater, BSc, PhD, MIMM
Mineral Resources Division
Institute of Geological Sciences

University College
FALMOUTH
Incorporating Dartington College of Arts

UNIVERSITY OF EXETER

Tremough Campus Library
STANDARD LOAN
This book may be borrowed for 28 days, subject to recall

Please return or renew this book on or before the due date, after which daily fines will be added to your account. Please check your issue slip or log onto 'my account' at http//voyager.falmouth.ac.uk

Library & Information Services
Tremough Campus, Penryn, Cornwall TR10 9EZ
01326 370441

http//library.falmouth.ac.uk or email library@falmouth.ac.uk

Mineral Resources

TELEPEN

London
Her Majesty's Stationery Office
1974

*Titles in the series*

| | |
|---|---|
| No 1 | Fluorspar |
| No 2 | Barium Minerals |
| No 3 | Fuller's Earth |
| No 4 | Sand and Gravel |
| No 5 | Tungsten |
| No 6 | Celestite |
| No 7 | Salt |
| No 8 | Sulphur |
| No 9 | Tin |

## Preface

The Mineral Resources Consultative Committee consists of representatives of interest Government Departments, and specialist advisers. It was set up in 1967 to keep present and future requirements for minerals under review and to identify problems associated with the availability, exploitation and use of mineral resources, both inland and offshore, having regard to competing demands on land use and other relevant factors.

Widespread and increasing interest in the mineral resources of the United Kingdom has led the Committee to undertake the collation of the factual information at present available about those minerals (other than fossil fuels) which are now being worked or which might be worked in this country. The Committee has produced a series of dossiers, each of which was circulated in draft to the relevant sectors of the minerals industry. They bring together in a convenient form, in respect of each of the minerals, data which had previously been scattered and not always readily available. These dossiers are now being published for general information.

## Acknowledgements

In the compilation of this report considerable reliance was placed upon the advice and information readily forthcoming from Mr K E Beer of the South-West England field staff, and from the Mineral Statistics and Economics Unit, of the Institute of Geological Sciences. Assistance was also received from the Department of Trade and Industry (including the Warren Spring Laboratory) and the Department of the Environment. The help of the following organisations is also gratefully acknowledged: Consolidated Gold Fields Limited; St Piran Mining Company Limited; the British Steel Corporation; the Cornish Chamber of Mines; London Tin Corporation; the Cornish Tin Smelting Company Limited; Consolidated Tin Smelters Limited; Capper Pass & Son Limited; the Tin Research Institute and the World Bureau of Metal Statistics. In particular, thanks are due to Mr J H Trounson of the Cornish Mining Development Association for his extensive comments on an early draft.

Metric units are employed throughout this document except where otherwise stated. In most cases this has necessitated the conversion of originally non-metric data. The units and conversion factors used are as follows:

| | | |
|---|---|---|
| millimetres | (mm) | = inches x 25.4 |
| metres | (m) | = feet x 0.3048 |
| kilometres | (km) | = miles x 1.609344 |
| hectares | (ha) | = acres x 0.404686 |
| kilogrammes | (kg) | = pounds x 0.45359237 |
| tonnes (1000 | kg) | = long tons x 1.01605 |

**Contents**

**Illustration**

## Summary

Tin, a metal principally used in the manufacture of tinplate, solder and non-ferrous alloys, is the only major, non-ferrous metal now mined in substantial quantities in the United Kingdom. Known resources of tin ore are confined entirely to a belt about 30 km wide and 150 km long between Land's End and Dartmoor in south-west England. In mid-1973 three underground mines were in operation, the old-established workings of South Crofty and Geevor, and Wheal Jane which was commissioned only in October, 1971. A fourth mine, Wheal Pendarves, started production in early 1972 but ceased in early 1973 when the operating company went into liquidation; however, it was announced in June 1973, that Pendarves ore would be milled at South Crofty after purchase of the mine by St Piran Mines Limited. In 1972 total domestic mine production was greatly augmented by output from Wheal Jane and amounted to 3,327 tonnes of tin-in-concentrate as compared with only 1,816 tonnes in 1971. Some concentrate is also produced by companies engaged in the treatment of old mine tailings and sands. In addition, nearly 5,000 tonnes of refined tin was recovered from waste and scrap in 1972, and an unknown, but probably larger, quantity is in circulation at any given time in secondary alloys which are re-used as such.

Primary and secondary refined tin from domestic sources amounted to more than 45 per cent of tin consumption in 1972, compared with less than 25 per cent in previous years. The United Kingdom imports tin, mainly in mineral concentrates, not only to meet demand, but also as feedstock for its smelting industry which exports substantial quantities of refined tin. The cost to the balance of payments of trade in unwrought and other forms of tin amounted to over £10.7 million in 1972 as compared with £17 million in 1971.

The International Tin Council, consisting of representatives of major producing and consuming countries, exerts a measure of influence on free market price movements through the operation of a buffer stock fund. Research sponsored by producing countries through the International Tin Research Council is aimed at maintaining and increasing the number and variety of outlets for tin, and is balanced by the efforts of consumers to reduce the quantity of tin used in each manufactured article. This applies in particular to tinplate, the tin coating on which has been gradually reduced in average thickness over many years without loss of performance.

The leading Free World producers of mined tin are Malaysia, Bolivia, Thailand and Indonesia, in that order, while the main consumers are the USA, Japan, the United Kingdom and the Federal Republic of Germany.

The economic benefits which would arise from an expanded domestic tin-mining industry include a saving of foreign exchange, lowered dependence on imports a source of concentrate for the smelting industry and improved employment opportunities in south-west England. Although little land is likely to be affected by mining, some prospects lie within areas generally considered to be of high landscape value.

## Introduction

Tin has been an important raw material since the early Bronze Age when it was first used by ancient Man to harden copper. It continues to be used in the manufacture of a variety of bronzes and other non-ferrous alloys but its dominant modern applications are in tinplate and solders. In these and numerous other uses, the metal's unique combination of properties have enabled it to compete effectively with alternatives, often on a basis of quality as well as cost, and the fact that it is generally used in small proportions in conjunction with other metals tends to overcome the disadvantages of its apparently high price.

Pure tin, which is white with a slight bluish tinge, is whiter than silver or zinc, although the surface colour depends upon the temperature of casting and the degree of polish. Its hardness is only 1.8 on Mohs' scale, and it can readily be cut with a knife. The tensile strength is poor and the melting point, 231.9°C at atmospheric pressure, is low. When exposed to extremes of cold, the metal undergoes an allotropic alteration to 'grey' tin. This starts to occur at temperatures below 18°C, but the process of alteration is too slow to be either noticeable or significant at normal temperatures, reaching a maximum only at -50°C.

Tin is superior in corrosion resistance to copper and nickel under normal atmospheric conditions, but a coating of tin on steel gives protection only where complete coverage is achieved. At an exposed contact of the two metals, corrosion of the steel is accelerated due to the electro-negative character of tin relative to iron.

Tin is not abundant in the Earth's crust, being present on average to the extent of only about 3 parts per million (as compared with, for example, zinc 80 ppm, copper 70 ppm, or uranium 2.5 ppm). The only important ore mineral of tin is cassiterite, a naturally occurring oxide, $SnO_2$. The high specific gravity of the mineral, between 6.4 and 7.1 according to variety, generally enables it to be concentrated mechanically by gravity methods such as jigging and tabling. Pure cassiterite contains 78.6 per cent tin but the best commercial concentrates average only about 70 per cent tin. The mineral is frequently brown or black, less often yellow, honey-coloured or red, and is more readily distinguishable from other minerals by its resinous lustre than by its colour. Owing to its high refractive index the rare transparent form is very attractive when cut as a gemstone but is somewhat soft and easily scratched. Other tin minerals, such as stannite (tin sulphide with some iron and copper) and cylindrite (sulphur of lead, antimony and tin), are not economically important although they have been recovered on a small scale, particularly in Bolivia.

Workable deposits of cassiterite are of two broad types. The more productive are those alluvial placer deposits of the kind exploited so successfully in Malaysia, Thailand and the Indonesian Archipelago. In these the cassiterite generally occurs as discrete grains separable from the containing sand by relatively simple gravity methods. The second major type of deposit is lode, or fissure vein, which is normally mined underground. Such lodes are the main sources of tin ore in Bolivia, a leading world producer, and in the United Kingdom.

Primary tin deposits are found only in the vicinity of granitic rocks, and the broad genetic connection between the two cannot be seriously questioned, although this does not mean that granites and neighbouring tin deposits are necessarily of the same age, or that the relationship is a simple one. Indeed, most granite masses are totally devoid of associated tin deposits; the regions of the world in which tin is found in workable quantities are well defined and relatively limited in extent.

Tin is traded in the form of concentrate and ingot, the former being a granular sand-like material composed largely of cassiterite grains. The practice of referring to tin concentrates by terms such as 'tin ore' or 'black tin' is misleading. The names 'wood tin', 'toads eye tin' and 'needle tin', amongst others, are descriptive of naturally-occurring forms of cassiterite. Terms such as 'Bangka Tin' and 'Straits Tin' are trade names which usually indicate the source of ingot metal or the smelting companies which produce it.

**Resources**

A region in Cornwall and Devon, about 150 km long and up to 30 km wide, is the only part of the United Kingdom in which tin minerals are known to occur in workable quantities and concentrations. The lodes and other deposits of this area are likely to remain the major, and probably the only, indigenous source of primary tin. Tin minerals have been recorded at Diebidale in Ross-shire, but no workable deposits have yet been proved. In 1927, the discovery of cassiterite in a foundation trench at Largs, Ayrshire, was reported but not confirmed.

In the south-west of England, tin-bearing lodes are associated with a major granite intrusion of Armorican age and the boundary of the mineralised area (Fig 1) closely approximates to the margins of the intrusion as they are delineated by both geological and geophysical data. The greater part of the mineralised region is composed of slate, shale, mudstone and greenstone, mainly Devonian in age, known locally by the collective term 'killas'. Prior to the intrusion of the granite, the sedimentary rocks and penecontemporaneous minor intrusives were folded and faulted on a generally ENE-WSW trend which evidently governed the form of the batholith when it was emplaced in late Carboniferous or early Permian times. Elevated parts of the granite intrusion are exposed on the high moors – Land's End, Carnmenellis, St Austell, Bodmin and Dartmoor – and in several smaller exposures and cusps.

In the western part of the peninsula the trend of the lodes is roughly ENE-WSW, while in the eastern part it is east-west. However, the lodes on the margins of the Land's End granite have a crudely radial distribution as compared with, for example, those to the south of St Austell Moor which are tangential to the granite mass. Some lodes coincide with a set of major joints in the granite. Most lodes dip steeply, at angles of 70° or more, although a few dip at 45° or less.

The region is not uniformly mineralised: on the contrary the tin-bearing veins are largely restricted to specific areas termed 'emanative centres' by H G Dines in 1956. Emanative centres may occur at the margins or near the centre of a granite boss, or in the killas at some distance, both laterally and vertically, from the granite. In some places the emanative centre appears to be located on a buried, or partly buried, cusp or ridge in the granite roof which may itself be mineralised, but the only evidence for the existence of an emanative centre is generally the presence of tin deposits. Structurally, therefore, only a general relationship between tin lodes and granite exposures is recognisable.

In detail the genetic connection between granite and tin is not at all clear. The position of the lodes may be influenced by the presence and attitude of quartz-porphyry dykes, locally referred to as 'elvans', coursing through the granite and killas of the region. Recently it has been suggested by J Dangerfield and J R Hawkes that tin deposition has resulted from the movement of volcanic waters along open fissures over prolonged geological periods, and that deposits are spatially dependent on the position of ancient erosion surfaces as well as the proximity of elvans. Although the specific means by which the minerals were deposited in the fissures remain unresolved, as also do the time relationships between mineral deposition and granite emplacement, the occurrence of successive zones of particular minerals around that the zoning is isothermal and caused by the deposition of a succession

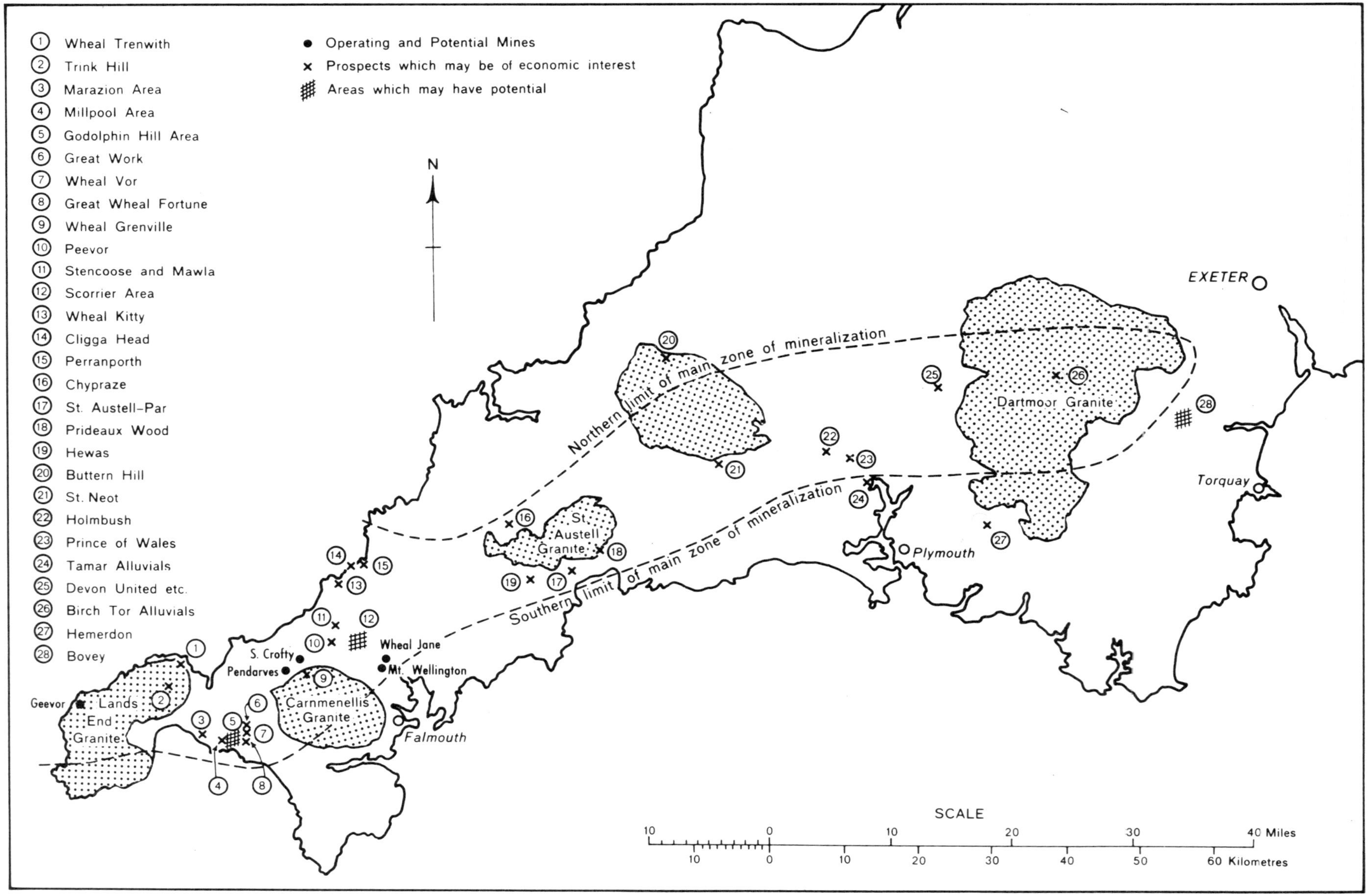


Fig 1 South-west England, showing approximate limits of tin mineralisation

of minerals at progressively falling temperature. Thus the metallic mineral nearest to the original heat source is cassiterite followed, further outwards or upwards, by minerals of tungsten, copper, zinc, lead and iron. However this zonal arrangement, apparent over the broad scale of Cornish geology, is not universally applicable although it is a feature of the strongest lode systems of Camborne and Land's End. The copper zone is not everywhere succeeded downwards by a tin zone, and iron lodes, for example, may occur within the granites.

The three main types of workable concentrations in south-west England are in fissure veins, vein swarms and alluvial (and eluvial) deposits. Fissure veins, which have yielded by far the greatest quantities of cassiterite, are planar bodies sometimes more than 1 km in length and depth but generally less than 1 m wide. Mining costs tend to be high because of the narrow stoping widths and the impracticability of using large excavating machines. Within the narrow confines of the fissure vein the ore generally occurs in shoots. Thus even in a rich lode only 50 per cent of the vein material may be payable and worth raising to the surface. Even in modern mining, vein material running much less than 1 per cent Sn is generally unworkable, although much depends on the width of the lode.

Vein swarms and stockworks are composed of ramifying or criss-crossing veinlets thickly distributed in more or less barren rock. In many places they are confined to narrow zones and are mined in the same way as fissure lodes. Occasionally they are laterally and vertically extensive and have been worked in open pits, e.g. at Carclaze and Hemerdon. Provided reserves lie near the surface and are sufficient to support a large throughput, costs are likely to be much lower per unit than those obtaining in lode mining, and workable grades may be considerably reduced.

Unconsolidated deposits of placer origin, particularly in buried river gravels, as well as eluvial deposits (residual concentrations), have been formed by the weathering of cassiterite-bearing veins during the Pleistocene. In these the cassiterite has been liberated, or partially liberated, from its matrix and may occur in discrete particles. The absence of blasting, raising and crushing costs enable such deposits to be worked at low cassiterite contents provided they are extensive.

Cassiterite has been produced in south-west England since the Bronze Age when the mineral was probably obtained entirely from gravels. Mining activity reached a peak in the second half of the 19th century (see Table 2) and the remains of mining works from this and earlier periods are scattered thickly over the landscape. Although many of the mines were abandoned for reasons other than exhaustion of ore, it must be emphasised that the working of Cornish lodes is more capital-intensive than it was a century ago and that many of the deposits mined at that time, if discovered at the present day, would not support a modern mining operation. The nature of deposits in the region is such that it is impracticable to make even an approximate estimate of the quantity of tin which remains to be recovered at given prices, for the steps leading to the discovery and assessment of deposits generally depend on the outcome of successive decisions involving progressively larger capital sums at total risk. In addition the cost of assessment work on a single lode system is likely to be high in relation to the information acquired, for drill sampling is only partially effective in determining the value of lodes, and recourse must normally be had to underground exploration by means of shafts and levels.

The region has been the subject of greater prospecting interest in the nineteen sixties than at any other period during this century, but it is significant that the only new mine so far brought into operation (Wheal Jane) is based upon a lode which is, for Cornwall, atypical in width, extent and mineralogy. The fissure veins upon which the past production of tin has largely depended are apparently economically submarginal and generally continue to be uncompetitive with deposits being exploited in other parts of the world. Furthermore, the thoroughness of mining in the past, and the success of prospecting during the 19th century in particular, seems to have left little margin for returns from the application of modern technology. Future mining in Cornwall and Devon may therefore be dependent upon the discovery of new deposits of the kind which can be worked on a large scale. Notable among those receiving attention are the vein swarms at Hemerdon and Cligga Head, which may be of this type.

The region may be conveniently subdivided by reference to the larger areas of exposed granite which are geographically prominent as extensive moorlands, namely, from west to east, Land's End, Carnmenellis, St Austell Moor, Bodmin Moor and Dartmoor (see Fig 1). The major mining districts associated with the *Land's End granite* are St Just and St Ives. On the north-west side of the granite a narrow strip of killas lies along the sea coast where some of the tin lodes, trending almost at right angles to the granite margin, have been worked under the sea. The tin zone, which is comparatively shallow within the granite onshore, pitches north-west seawards. Prominent among the mines operating in this area during the last century were Geevor, Levant and Botallack, the first of which is still in production. The Levant mine, the 300-fathom level of which runs for about 1½ km out to sea, closed in 1930, and recent work aimed at reinstating operation of the Levant lodes, involving the sealing of a breach between old workings and the sea, has proved disappointing. While the seawards extensions of the lodes may be favourable targets for prospecting at depth, the capital and operating costs of mining at progressively deeper levels away from a shaft on land are likely to be discouraging factors. In addition to the large mines on the coast, numerous relatively small workings which produced a few tens of tonnes of ore were operated within the granite. Alluvial deposits on the moor, particularly in the Morvah-Sancreed area, were worked in ancient and later times, and it is unlikely that any remain payable.

On the north-east side of the Land's End granite near St Ives, tin was mined in the granite country in such workings as Wheal Reeth. The amount of tin recovered from the coastal strip of killas was small in comparison, and the lodes, trending roughly perpendicular to the granite margin, were not generally pursued seawards. The workings of Wheal Trenwith, parts of which underlie the town of St Ives, were originally developed for copper, but the possibility of finding tin shoots is reported to be encouraging as tin was discovered in 1922 in the lower levels, 180 m below adit, before the operating company St Ives Consols Mines Limited, was forced into liquidation. The lodes on the western and southern flanks of Trink Hill have also been regarded as targets for investigation, although prospecting by Baltrink Tin Limited in the period 1966 to 1968 indicated that eastward extension of the Giew Main Lode is limited by a sandfilled crosscourse. Attempts by this company to carry its investigations further by increasing the area of its concession were prevented by difficulties in establishing ownership of mineral rights.

Between the Land's End and Carnmenellis granites are the extensive tracts of killas of the Gwinnear and Mount's Bay districts. In Gwinnear, to the south-east of Hayle, where the killas is extensively intruded by greenstones and elvans, irregularly mineralised lodes were worked principally for copper. Although some tin was recovered, the area was not an important tin producer. By contrast, in the Mount's Bay district to the south, some of the most important tin mines in Cornwall occur in the proximity of the Godolphin granite, a small outcrop some 3 km across and 5 km long. The bulk of tin production came from three groups of mines: those around Millpool, to the west of the Godolphin granite, those near Great Work on the eastern edge of the granite and those associated with Wheal Vor and Wheal Metal in the killas country between the Godolphin and the Carnmenellis granites. In the Marazion district, north of St Michael's Mount, a tract of ground about 1 km wide between the Marazion Marshes and Goldsithney contains a number of old copper mines including Wheal Hampton, Tregurtha Downs and Owen Vean. Some tin was recovered at the deepest levels of Wheal Hampton before the mine was abandoned in 1914. Sporadic drill sampling carried out in the late nineteen-sixties is reported to have been discouraging, but further investigation may be justifiable in future. Similarly, in the Millpool area, little is known of the distribution of tin beneath the existing shallow workings. In some places it occurs in lodes separate from those of copper, while in others it has been found below the copper zone as at Wheal Lewis and Millpool mine. The extensive elvan dykes in the area may have influenced the deposition of tin.

Among the more interesting tin mines in the district are Great Work and Wheal Reeth, both of which have been worked to over 300 m below adit level. Depth extensions of the lodes may prove rewarding targets for investigation, and their proximity to an emanative centre suggests the possibility of finding 'blind' deposits.

In the group of mines near Wheal Vor, the deposit at Great Wheal Fortune, formed by the occurrence of numerous cassiterite veinlets in highly fractured slates, may be workable by open pit if sufficiently extensive. Values obtained in former surface working were reported to have averaged about 4 kg per tonne.

The lodes associated with the *Carnmenellis* granite, and the neighbouring bosses of Carn Brea and Carn Marth, include some which have been among the most productive in Cornwall. The highly mineralised belt which skirts the northern margin of the Carnmenellis granite can be regarded as the centre of the tin mining industry, embracing the well known mines of Dolcoath, East Pool and South Crofty, the last of which remains in production. On the southern edge of Carn Brea the Great Flat Lode is a large, gently-dipping structure approximately following the contact of the granite and a belt of killas which marks a depression in the roof of the batholith north of Carnmenellis. It is intersected by many steeper lodes at depth. North of Carn Brea the trend of the lodes is parallel to concealed ridges in the roof of the batholith. Almost all the mines in this area were originally on copper lodes which became tin-bearing at depth.

Situated in a line along the northern edge of the Carnmenellis Granite, the Grenville, South Condurrow and Basset mines worked the Great Flat Lode in preference to vertical lodes, the continuity of which has been established by recent drilling. The lodes are often rich, but tend to be narrow, and although extensive ground remains to be developed, the question as to whether or not sufficient ore exists for profitable mining can be resolved only by further investigation. However, the area also offers the possibility of another big structure parallel to, and 60 to 100 m below, the Great Flat Lode. This 'twin' lode has been reported in a few boreholes but has not been followed up. Unstoped parts of the Great Flat Lode, averaging about 7 kg of cassiterite per tonne, have been intersected within the Basset area.

To the north-east of Carnmenellis, the Scorrier mining district embraces the Carn Marth granite and the killas country on its eastern and northern flanks. The workings in the area, mainly for copper, are generally shallower than 200 m below adit level, and many date back to the 17th century. Future development of tin mining in the district must depend upon whether the copper zones are succeeded downward by those of tin, in analogy with the Camborne district to the south-west. Further investigations aimed at locating tin deposits might be worth conducting in the north-east of the area, under such mines as Wheal Busy, Creegbrawse, Hallenbeagle, Killifreth and Unity Wood. In the nineteen-sixties drilling was carried out at Peevor mine, about 1½ km west of Scorrier, by Camborne Tin Limited. Later, the Barcas Mining Company embarked on a programme of underground exploration which was abandoned after severe flooding. The property might repay examination in conjunction with neighbouring mines, notably Little North Downs, West Peevor, Treleigh Consols, Cardew and Prussia.

There is geological evidence to suggest that a shallow subterranean ridge of granite extends from the Carn Marth mass to St Agnes Head on the north coast. In the intervening belt of killas an old copper working, Stencoose and Mawla United, is reputed to have encountered tin-ore at depth before being flooded. The discovery of tin mineralisation at the surface about 1½ km east of Stencoose suggests that extension of the lodes may be considerable.

The St Agnes district comprises the country on the north coast of Cornwall around the small granite bosses of St Agnes and Cligga Head. Near the granites and, more notably, between them, the lodes are largely tin-bearing and trend ENE-WSW. Only a few lodes have been worked in the granite. The chief tin-producing deposits are gently-dipping veins which were worked in Wheal Kitty, Polberro and Blue Hills mines. At Cligga Head the granite is traversed by closely spaced tin-tungsten-bearing quartz veins which were investigated prior to the Second World War when it was considered that the veinlets were grouped in three near vertical zones between 7 and 14 m wide, averaging about 5½ kg per tonne of mixed wolframite and cassiterite. These were mined by Rhodesian Mines Trust Limited until 1945. In mid-1973 the deposit was being re-assessed by International Mines Services Limited. To the south and east of the Cligga vein-swarm the lodes in the killas gave rise to the copper mining industry of Perranporth which ceased operations in the mid-nineteenth century. The deepest mine is only 230 m below sea level, and it is reasonable to suppose that the lodes may become tin-bearing at depth. A shaft in the Cligga granite would permit the crosscutting not only of the lodes in the Perranporth mines, but also any parallel lodes occurring to seaward and the possible depth extensions of the Cligga veinlets which might coalesce downwards into single lodes.

Eastwards, the next important group of mines is that in and near *St Austell Moor.* In this area the killas consists of slate interbedded with calcareous rocks which are altered to calc-silicate hornfelses in the aureole of the granite. The granite itself is widely kaolinised and traversed by numerous veinlets of quartz and tourmaline. Where these contain cassiterite, and are closely spaced, they have been mass-mined, often with china clay, as in the large open pit of Carclaze. Elsewhere cassiterite has been recovered from time to time as a byproduct of china clay working. The most productive mines in the St Austell area lay to the south-east and south-west of the granite mass where the lodes trend parallel to the granite margin and dip towards it. The tin and copper mines of the south-eastern group occur within an area extending from St Austell to Par and inland from the coast for about 2½ km. At the western end of the belt the mines yielded tin, whereas those in the east produced copper. Par Consols, at the eastern end, was abandoned in 1869 after exhaustion of the copper shoots, although tin was being encountered at depths of 500 metres. The Wheal Eliza group of mines to the west produced over 11,000 tonnes of tin concentrate from the same run of lodes. There may therefore be further resources of tin below the copper zone in the northern parts of Par Consols. Further west, around Sticker, the mineralised, crushed slates of St Austell Consols, Great Dowgas and Polgooth may also offer limited possibilities for future tin production.

The upper levels of the Hewas Mine, in the group lying to the south-west of the St Austell granite, are evidently exhausted. This mine produced about 24 tonnes of concentrate monthly for six years after being reopened in 1919, output being based on a new set of four lodes mined to a maximum depth of only about 100 m. The older workings are also shallow, the Main Lode having been developed to only about 175 m below surface. The property is geologically well situated and offers the possibility of new discoveries, as well as depth extensions in worked lodes.

On the east side of the granite mass is the group of mines around Fowey Consols. A property near Prideaux Wood on the western extension of Fowey Consols has not been mined, although it is favourably situated with respect to the contact of granite and killas. The yield from the lodes of Prideaux Wood mines, which were worked to a depth of only 60 m, should now be capable of improvement although, in current circumstances, the available ore in both setts is unlikely to be sufficient to warrant mining. At Chypraze, about 1½ km south-east of St Enoder, gravels on the wide alluvial tract of one of the tributaries of the Tresillian River might offer opportunity for small working.

Although the killas in the neighbourhood of the *Bodmin Moor* granite is host to a great number of mineralised veins, many contain only low temperature minerals of lead, zinc or iron. The most important tin mines were those of the Phoenix and Caradon groups mainly within the granite country in the south-east of the area. Near St Neot, trials at St Neot and Kilham mines showed that a gently dipping polymetallic lode extends over a strike length of about 1½ km. This has been superficially examined at each end to reveal a wide complex lode with patchy cassiterite. No work has been done below adit. On Goonzion Downs mineralised elvans and several lodes have been trenched, but little underground work has been attempted.

On the eastern margin of the granite near Bowithick village there are large stretches of alluvium which might repay investigation. Alluvial deposits in the valley between Bray Down and Buttern Hill have been worked but contain only patchy cassiterite and are of limited extent. In the valleys west and south of Five Lanes, Altarnun, extensive narrow stretches of tin-tungsten-bearing alluvium, most of which have been worked in the past, still contain worthwhile values.

Between Bodmin Moor and Dartmoor the twin granite cusps of Kit Hill and Gunnislake are situated within an important belt of mineralisation, 6.5 km wide, extending westwards from the edge of Dartmoor. Tin values tend to be patchy and the lodes generally contain a relatively high proportion of wolframite. The most productive tin mines were Drakewalls and Prince of Wales, both lying in the slaty killas to the south of the Gunnislake boss where there may still be scope for finding new lodes. A group of mines known as Callington United, formerly the workings of Holmbush, Kelly Bray and Redmoor, is situated on the western flank of the Kit Hill granite. These produced copper, arsenic, tungsten and lead as well as tin, the deepest levels being only 400 m below surface. Detailed reports made at the time of closure in 1888 indicate that the Holmbush and Kelly Bray lodes were becoming tin-bearing with depth and suggested that Redmoor might become a substantial producer of tin. About 4 km ENE of the Prince of Wales mine, the Ding Dong mine was tested by Non-ferrous Minerals Development Limited during the Second World War with encouraging results, although the quantity of ore available was reported to be limited. Foundation investigations for a railway bridge across the Tamar at Calstock in 1907-08 revealed cassiterite and wolframite in alluvium about 40 m deep. Subsequent drilling in the area indicated a grade of over 2 kg of cassiterite and wolframite per tonne, suggesting that the alluvium in the Tamar River flats below Gunnislake might be economically workable. However, the area is noted for its considerable beauty, and no mining has ever taken place.

The most easterly of the granites of the mineralised region, *Dartmoor*, is perhaps more notable for metals other than tin, although tin was mined at several centres, notably Devon United and Friendship mines to the west of the granite near Marytavy, and at sites in the central and southern parts of the moor. The Birch Tor alluvial deposits, situated in central Dartmoor, were worked in the remote past and found to be up to 10 m deep when sampled in 1930. They are composed of a mixture of sand and rock fragments, the latter containing a high proportion of lode material. Rough sampling indicated values of between 2 and 3 kg/tonne, the sand containing up to 3.5 kg/tonne and the rock fragments from 2 to 5 kg/tonne. About 160 hectares of alluvial deposits are present in three adjacent valleys and may offer some opportunity of return with the added advantage that, if carefully executed, the work of recovering the tin could also enhance the amenity value of the area and make the old openworks safer.

To the south of Dartmoor a small granite boss containing a swarm of closely-spaced greisenised tungsten and tin-bearing quartz veins is exposed on the hill of Hemerdon Ball. There are two sets of veins, both dipping at moderate angles, one set trending roughly ENE-WSW and containing cassiterite and wolframite, the other trending NNE-SSW and bearing wolframite only. Available information suggests that the body extracted **en masse** would grade 0.024 per cent tin and 0.093 per cent tungsten trioxide. To mine such low grade material successfully would require large throughputs, probably exceeding 5,000 tonnes per day, as compared with a total known reserve of only 5 million tonnes. Further work will therefore almost certainly be aimed at determining the full extent of the mineralised body. A brief summary of the history of the property is given in Mineral Dossier No 5, 'Tungsten'.

On the south-east and the south-west margins of the Bovey clay basin to the east of Dartmoor a series of chert and flint sands and gravels have yielded fine-grained cassiterite in panning samples from shallow surface pits. Values vary but appear to average 0.2 kg cassiterite per cu m on the basis of a limited number of samples. Though the grade is low, tonnage is potentially very large and a degree of concentration could be achieved by simple screening.

As the long axis of the mineralised area corresponds approximately with that of the Cornish peninsula, it might be supposed that good prospects for submarine placers exist in the near-shore regions of the English Channel and Irish Sea. However, work carried out to date off Cornwall and Devon indicates that natural placers are unlikely to be extensive and that values are probably confined to old mine tailings in the upper metre or so of bottom sediment where the sea bed has not been swept by current action. Trials in St Ives Bay were abandoned in 1967 because a large proportion of the cassiterite was found to be locked in heavy silicate minerals, a factor which, together with the fine grain of the cassiterite and the difficulty of beneficiating the sands aboard ship, resulted in very low recoveries There may thus be only limited scope for the recovery of cassiterite by marine dredging, for large scale operations would depend for success on the presence of extensive reserves in which a high proportion of the mineral is liberated. However, since 1967 the Marine Mining Corporation has undertaken large scale reconnaissance of bottom sediments off the coast and is developing a technique for dredging in shallow water with a large tide range.

Advances in mineral dressing techniques, the acceptability of lower concentrate grades and increases in the price of tin have made it possible in certain conditions to recover cassiterite profitably from old mine tailings. Several tailings treatment plants are generally in operation in Cornwall at any given time. Production usually amounts to a few tens of tonnes annually from material grading less than 0.1 per cent tin.

In an industrialised country, such as the United Kingdom, resources of tin include scrap metal and alloys in various forms. At any given time these form substantial reserves additional to available reserves of the metal in the ground and in old mine dumps. Indeed, the contribution of secondary metal to domestic consumption, in terms of secondary refined tin and tin-in-alloy, has, in the last twenty years at least, exceeded that of domestic mined tin. Refined tin, recovered largely from tinplate scrap, and tin-in-alloy (bronzes, Babbitt, type metal) recirculating indefinitely as alloys of particular composition and application, form an important national resource. Figures for the recovery of tin-in-alloy are not published and would be difficult to determine, although it may be significant that, in the USA, recovery of secondary tin-in-alloy amounts to about nine times that of secondary refined tin. In view of the large proportion of tin used in tinplate, it is unfortunate that recovery of the metal on a large scale from scrap food cans in domestic refuse is not currently economic. The resulting wastage of tin is reported to be as much as 7,000 tonnes per year, worth well over £10 million at mid-1973 prices.

## Land use

The area of land affected by current mining for tin in south-west England is small, for, although existing planning permissions for the working of tin and associated minerals apply to some 3,640 hectares, these permissions cover the whole of each mining sett, only a small part of the surface of which is used for buildings and dumps. Of the three existing mines, Wheal Jane is sited in an area made derelict by past working; South Crofty adjoins the built-up area of Pool; and Geevor is surrounded by rough pasture. Below Pool, in the Red River valley, there is a concentration of alluvial workings.

Although centuries of uncontrolled mining, mainly for copper, have given rise to serious dereliction, there is little possibility that future workings would substantially disfigure the landscape. Current planning procedures provide that permissions can include such requirements as the prior approval of the local planning authority before the erection of plant and building, the prevention of subsidence and the clearance of surface installations on completion of mining. The appearance of any new installation would also be regulated by planning control.

Although the proportion of land used for mining is only one of the factors to be taken into account in a consideration of the future effect of mining on amenity and agriculture, it is notable that recent studies by the Cornish Mining Development Association indicated that ores of all the base metals might be workable beneath 36 areas covering in all about 104 sq km or only 2.92 per cent of the land surface of Cornwall. However, in the unlikely event that mines in all these areas came into operation simultaneously, the surface area occupied by mining works and waste dumps would amount to less than 0.1 per cent of the total land area.

Normal lode mining, involving the stoping of ores at depth from narrow, steeply inclined lodes, does not generally give rise to problems of subsidence. Collapse around old shafts and the presence of open shafts in old workings are obvious hazards and are recognised as such by the Local Authorities. Cornwall County Council has undertaken the fencing of those known to be dangerous.

The despoliation caused by past mining is a major factor influencing public reaction to further mining ventures in south-west England. The area in Cornwall categorised as derelict land amounts to about 65 sq km (excluding china clay tips and workings), and represents the largest such area for any county authority in England and Wales. Cornwall County Council have returned the following data (in hectares) on derelict land as of December, 1971, which show a slight improvement on previous years:

| *Spoil Heaps* | *Excavations and pits* | *Other forms e.g. Buildings* | *Total* | *Total justifying treatment* |
|---|---|---|---|---|
| 5,496 ha | 322 ha | 642 ha | 6,460 ha | 694 ha |

Dereliction is accentuated in some places by the effect of metallic compounds (including arsenides) derived from mine waste, which may contaminate the soil and inhibit re-vegetation.

The conditions imposed on new ventures to prevent abuse of landscape and amenity is likely to add substantially to costs. Capital charges on new mines may be significantly increased by the necessity to purchase land for slimes ponds in order to obviate pumping of slimes into natural waterways, and by conditions regulating siting, external appearance, screening, restoration and after use. Much of the expense will fall upon the venture at its most vulnerable stage, i.e. during pre-production development, before generation of cash flow. The industry is not necessarily able to pass such extra costs on to the consumer as its revenue is determined largely by prices on international markets.

## Uses

Because of its low strength and high cost, tin is seldom used by itself but normally finds application in the form of coatings on other metals and as a component of alloys or compounds. The metal usually forms only a small, although often vital, part of manufactured articles and its importance is therefore not generally recognised. The ubiquitous tin can, for example, is largely composed of mild steel, the protective layer of tin on which is generally less than 0.0004 mm thick. Similarly, the presence of tin in the cylinder block and main bearings of a motor car, or in the soldered contacts of electronic equipment, is an insignificant feature of these articles in terms of weight and relative costs.

The usefulness of tin stems from its unique combination of properties. Perhaps the most important among these are its proven non-toxicity and resistance to corrosion. Its low melting point is also an advantage in many applications, as is its ability, when molten, to wet and remain firmly coated upon other metals. In addition it alloys readily, it can be deposited as an effective layer by electroplating methods, it forms a wide range of compounds, both organic and inorganic, and possesses a low coefficient of friction.

The two major uses of tin are in tinplate and solder which together account for two-thirds of world tin consumption (44 per cent and 20 per cent respectively). The other main uses are in bronzes, bearing metals, tinning (e.g. of copper and steel wire), organotins (tin-based organic compounds) and inorganic tin compounds. The pattern of consumption varies considerably between one country and another. In the United Kingdom the quantity of tin consumed in alloys greatly exceeds that used for manufacturing solder.

Although methods of applying relatively thin layers of tin to other metals have been known and used for hundreds of years, their large scale application in the preservation of food dates only from the beginning of this century. Technological advances, in particular the change from hot-dipping to electrolytic tinning which began in Germany in the nineteen-thirties, have ensured the competitive position of tin plate, now a versatile material produced to a range of specifications and in various coating thicknesses. The principal advantages of tin plate are its strength, weldability, resistance to corrosion and non-toxicity. The ability of the tin coating to conform to the bending or drawing of the steel base enables the tin plate to be manipulated into complex shapes without the risk of failure of the coating. It is used for packaging oils, polishes, prints, cosmetics and other preparations, as well as food, and for making bottle tops, kitchen utensils and light engineering parts.

Soft solders are alloys of tin and lead. Those containing a high proportion of tin (up to 65 per cent) are the more expensive and generally the most effective. The electrical machinery and electronic equipment industries, in particular, are dependent on high-tin solders for reliable contacts. Solders with lower tin content are used in the manufacture and maintenance of vehicle radiators and other heat exchangers, in the repair of motor car bodies and in plumbing.

The prehistoric discovery that the addition of tin had the effect of hardening copper led to the manufacture of the earliest metal tools. Modern bronzes containing up to 25 per cent tin include phosphor bronzes, gunmetals, bearing metals and bell metal. The usual proportion in gunmetal, for example, is 90 per cent copper and 10 per cent tin. The various bronzes have a wide range of uses including the manufacture of valves, pumps, bearings, gears, springs and condenser tubes.

White anti-friction metals generally contain over 90 per cent tin, but due to lack of fatigue strength these have been largely superseded as bearing metals in highly-stressed engines, originally by copper-lead bearings and more recently by aluminium-tin alloys containing about 20 per cent tin. Printers' alloys or type-metals are alloys of lead, antimony and tin, with between 5 and 20 per cent tin. A range of lead-tin and tin-lead alloys known as 'Babbitt' have had wide application. Pewter, an alloy which for centuries has been fashioned into tableware and decorative vessels and plaques, is a high-tin alloy containing about 20 per cent lead, although numerous compositions have been employed. Britannia metal, a modern pewter, is a tin-antimony alloy often containing small amounts of copper and zinc.

Many forms of polyvinyl chloride contain up to 1 per cent by weight organotin compounds for the purpose of decolorising or stabilising of colour during manufacture. In the nineteen-fifties, dibutyltin stabilisers were perfected and applied in PVC where clarity and rigidity were required, but with the development of cheaper calcium-zinc and barium-cadmium stabilisers, their use became largely confined to the minor proportion of PVC in which maximum clarity and rigidity were specified. However, although the rate of increase in demand for dibutyltin has been much lower than that for PVC there has been considerable growth in demand for dioctyltin stabilisers which are suitable in the manufacture of clear, colourless PVC foils and bottles.

Organotins have passed into established use in numerous other applications. These include triphenyltin as a fungicide for potatoes, tricyclohexyltin hydroxide as a miticide and tributyltin oxide as a biocide. Other tributyltin formulations are used in anti-fouling paints and wood preservatives, as a slimicide in paper manufacture (to replace the toxic organomercury) and as a control for bilharzia and rodents.

The inorganic compounds of tin include stannous fluoride used as an additive to toothpastes, tin oxide for ceramic glazes and vitreous enamel opacifiers and tin chloride for electroplating salts, soap perfume stabilisers and as a weighting agent for silk.

Recent new uses include the addition of small quantities of tin to cast irons to give uniformity of hardness and wear resistance, the application of molten tin to the manufacture of 'float glass', and the alloying of tin with niobium for the manufacture of superconducting tape.

**Specifications**

The lowest-value ex-mine product which is readily marketable is tin concentrate. This may be high grade, containing up to 75 per cent tin, or lower grade, containing as little as 15 per cent, and is liable to contain a progressively greater proportion of deleterious impurities with decreasing tin content. The variety of possible concentrate compositions is so great that no standardisation is possible. Each concentrate must be sampled, analysed and subjected to valuation by bargaining or upon a previously agreed schedule.

Ingot tin, the normal product of the tin-smelting industry, is the subject of British Standard 3252, 1960. This covers the chemical compositions of three grades of ingot and the general requirements relating to inspection and testing.

Other British Standards relating to tin products are as follows:

| | |
|---|---|
| BS 219, 1959 | (with amendments PD 4843, and PD 5087, 1963) Soft Solders. Covers the compositions of eight grades of non-antimonial solders, six grades of antimonial solder for general use, and three grades of solder for service at higher temperatures. |
| AU 90, 1965 | Soft solders for automobile use. |
| BS 1468, 1967 | Tin anodes and tin salts for electroplating (Amended October 1968). Deals with the physical form, chemical composition and impurities for tin anodes, stannous sulphate, sodium stannate and potassium stannate. Methods of analysis are given. |
| BS 1872, 1964 | Electroplated coatings of tin. Specifies requirements for significant surface appearance, heat treatment and sampling of tin coatings on fabricated articles of iron, steel, nickel alloys, copper and copper alloys. Requirements and methods of test are also given for thickness, adhesion and solderability. |
| BS 2920, 1969 | Cold-reduced tinplate and blackplate. Covers blackplate, hot dipped tinplate, electrolytically equally coated and differentially coated tinplate in terms of coating weights, tempers, sampling and testing procedures, tolerances, packaging, marketing, surface finish and trade terms. |
| BS 3338 | Methods of sampling and analysis of tin and tin alloys; Parts 1 to 6, 1961 deal with the sampling of, and determination of non-ferrous metals in, ingot tin. Parts 7 and 8, 1961, are concerned with the determination of silver and bismuth in tin alloys. Parts 10 to 12, 1961, parts 9 and 13 to 17, 1965 and parts 18 and 19, 1966, deal with sampling of ingot tin and tin alloys, and the assay for various elements in them. |
| BS 3597, 1963 | Electroplated coatings of 65/35 tin-nickel alloy. |

BS 3788, 1964 Tin coated finish for culinary utensils. Deals with the chemical composition of such coatings and the general surface finish.

In addition, there is a wide range of British Standard specifications for bronzes for various purposes, and BS 3332, 1961, gives the chemical composition of nine white metal bearing alloys in ingot form.

The specifications and methods embodied in British Standards are a small fraction of procedures and compositions which are widely published and understood. In traditional applications, such as pewter, tin bronzes and white metals, numerous alloying compositions are in use, and one leading authority lists over one thousand alloys containing tin, many of which, with names such as Mock Silver, Zeppelin Alloy, Laderig's Speculum, Ashberry Metal, probably have only limited application, if any, in modern industry.

**Price**

The prices paid for tin concentrates are determined by the ruling London Metal Exchange price of the metal and by the smelters' treatment charges (in effect, the price of smelting) which are a matter for negotiation between the owner of the concentrate and the smelter. Details of these charges for tin concentrates are published twice weekly in the 'Metal Bulletin'; in mid-1973 they were as follows:

70-75 per cent Sn (less 1 unit) £25 per tonne
40-65 per cent Sn (less 1-1.6 units) £51-£56 per tonne
20-30 per cent Sn (including deduction) £95-£100 per tonne

Thus it would cost the owner of a 75 per cent Sn concentrate £25 plus 1 unit of tin (10 kg) to have one tonne of concentrate smelted. For each tonne of concentrate presented to the smelter he would receive in return 0.74 tonne of metal (i.e. 750 less 10 kg). Penalties according to agreed schedules are applied to concentrates containing impurities.

In practice, concentrates produced in Cornwall are purchased outright by smelting companies on the basis of London Metal Exchange quotations averaged over an agreed period. From this valuation are deducted the smelting charge, such penalties as may be applicable and the unitage deduction where provided for in the smelting terms.

The price of tin metal is determined largely by market forces operating through the London Metal Exchange and the Penang market. The United Kingdom is a member of the Fourth International Tin Agreement which came into force in July 1971, and which is administered by the International Tin Council. The two principal objectives of the Tin Agreement are to prevent excessive fluctuations in the price of tin by adjusting world production to consumption, and to ensure adequate supplies of tin at prices which are fair both to consumers and producers.

The means available to the Council for pursuing these objectives is the maintenance of a Buffer Stock financed by producer members. By means of purchases and disposals on the London Metal Exchange and other markets the Buffer Stock Manager attempts to keep the price of tin within the limits determined from time to time by the Council. The operating ranges of the Buffer Stock lie between a floor price (at which the Manager must buy) and a ceiling price (at which he must sell). Between these limits there are three ranges: the Manager may operate in the upper range as a net seller and in the lower range as a net buyer. Unless specially directed by the Council in unusual circumstances (e.g. when export controls are in force) the Manager does not operate in the middle range.

An important feature of the London Metal Exchange is that it is a physical market, enabling producers to offer to sell their products at any time and in any quantity, and providing fabricators and manufacturers with the facility to purchase supplies of tin metal at all times. The Exchange also offers producers, consumers, stockists and merchants the facility for hedging operations. Penang and New York however, are purely markets determining prices at which buyers purchase and take delivery. Average annual LME tin prices for the period 1953-1972 are given in Table 1.

**Table 1 London Exchange: tin prices, 1953-72**

*Annual average price per tonne*

| | £ | | £ |
|---|---|---|---|
| 1953 | 720.1 | 1963 | 895.3 |
| 1954 | 708.0 | 1964 | 1,219.8 |
| 1955 | 728.4 | 1965 | 1,390.4 |
| 1956 | 775.3 | 1966 | 1,275.3 |
| 1957 | 742.9 | 1967 | 1,209.4 |
| 1958 | 734.9 | 1968 | 1,323.3 |
| 1959 | 773.0 | 1969 | 1,428.4 |
| 1960 | 784.0 | 1970 | 1,529.5 |
| 1961 | 874.6 | 1971 | 1,437.4 |
| 1962 | 882.3 | 1972 | 1,505.9 |

*Source:* International Tin Council

The widespread industrial use of tin, despite its high price, is based largely upon the fact that only a small proportion of the value of end-products containing the metal is attributable to its cost. For example, various thickness of tin are used in tinplate manufacture but a commonly used coating (referred to technically as 0.25 lb per basis box) is only 0.00038 mm thick on each face, equal to 0.0028 kg per sq m. Thus in a can composed of 0.05 sq m of such plate the element of cost arising from the tin content would amount to about 0.035 pence. Similarly in electronic and other electrical installations the increment in the price arising from the cost of tin-in-solder is trivial.

The cost of transport is an insignificant factor in the price of tin metal. The cost of moving concentrates from mines in Cornwall to the Liverpool smelter in mid-1973 was about £3.75 per tonne, and that for transporting metal by road from Liverpool to principal consumers in South Wales amounted to a little over £3.00 including insurance.

## Technology

In south-west England cassiterite has mainly been won from fissure veins. Normal mining practice is to crosscut to the lodes from a carefully sited shaft and to drive along the veins from the crosscut. The lode material is usually brought down into the drives by overhand stoping and trammed along the haulage ways to the shaft. In some areas cassiterite has been recovered from alluvial deposits by excavation and treatment of the sands, and from stockworks by open-pit quarrying.

Depending on the depth of working, the shaft capacity, width of the lodes and other factors, the cost of mining and hoisting ore in a Cornish lode mine is currently (mid-1973) about £5 to £7 per tonne.

In lode mining, cut-off grade depends on mining conditions and the metallurgical complexity of the ore. Generally, average millhead grade in Cornwall is about 1 per cent tin. Alluvial deposits, in the most favourable conditions such as those obtaining in parts of Malaysia, can be worked profitably by dredge down to 0.1 kg per cu m (cassiterite) which is roughly 0.0053 per cent tin. In tin-bearing sediments which might occur off the Cornish coast, much higher grades would be necessary for profitable working, particularly if they were derived from old mine tailings in which cassiterite is incompletely liberated.

Tin ore is normally processed by physical concentration of the tin-bearing mineral to form a concentrate which is smelted to win the metal. The complexity of processing depends upon the size of the mineral particles, the nature of the gangue and the presence of such deleterious impurities as antimony, bismuth, arsenic, copper, lead and zinc. Milling costs are between £1 and £1.75 per tonne depending on the complexity of the ore.

As cassiterite has a high specific gravity it is normally concentrated by gravity methods using jigs, sluices, tables and spirals. Electromagnetic and electrostatic separators are often employed in the final part of the process to free the cassiterite from other heavy minerals.

The steps in the milling of a moderately complex lode might be as follows:

(a) Primary crushing, washing and screening.

(b) Heavy-media separation and rejection of low density rock fragments.

(c) Secondary comminution.

(d) Hydraulic classification and gravitational separation of mineral assemblage on vibrating tables with possible recycling of middlings after further comminution.

(e) Slimes treatment and recovery of fine values.

(f) Removal of sulphides by flotation.

(g) Final concentration and magnetic separation of such minerals as magnetite, siderite and wolframite.

(h) Assaying, weighing and packaging of final concentrates for the smelter.

Direct flotation of cassiterite from slime fractions is now considered an economic process which is being adopted commercially. At the Wheal Jane operation in Cornwall, improved flotation techniques, developed in conjunction with the Warren Spring Laboratory, are being used to deal with comparatively fine grained cassiterite.

Concentrates generally contain from 60 to 70 per cent metallic tin, but those containing from 40 to 60 per cent are also readily marketable. Concentrates with less than 30 per cent tin are acceptable to specialised smelters. With the further development of the sulphide fuming process, concentrates with as low as 5 per cent may eventually find a ready market.

There are several possible methods of extracting metal from concentrate. Commercially the most important of these is by reduction in the reverberatory furnace with coke or anthracite and limestone after roasting to expel arsenic and sulphur. Smelting is carried out in two stages. In the first, moderate temperature and reducing conditions are applied to produce comparatively pure tin. This process, however, leaves a very high proportion of tin in the slag which is therefore refurnaced at higher temperatures with more reducing agent bringing down more tin, in particular the globules or 'prills' of the metal entrapped in the primary slag. The metal produced in the second stage is less pure than that from the primary smelt as iron is also reduced at the higher temperatures and combines with a proportion of the tin to form an alloy known as 'hardhead'. Variations of this procedure are used, including for example, fuming of the primary slag to produce a clean tin oxide fume which is smelted in a second reverberatory furnace.

In refining, the removal of solid impurities, notably iron, is achieved by raising the temperature of the impure tin to a point just above that at which it melts. The liquid tin is permitted to drain off, or may be separated by filtration from its solid impurities. Arsenic and antimony are precipitated in the form of $AlAs$ and $AlSb$ by adding aluminium. Other impurities are drossed by other additives. These processes come under the heading of 'liquation' or 'dry refining'. An alternative process now in commercial use is vacuum refining, effective in removing lead and bismuth.

Although electrolytic refining produces a higher purity metal, the capital investment necessary for equipment and the stock of crude metal required per throughput unit are such that the process is used largely for metal containing impurities of some value, or for the byproducts of dry refining.

The basis upon which smelter costs are determined is referred to on page 18. It is worth noting the rapid increase in these costs per tonne of metal as concentrate grade decreases. A 70 per cent Sn concentrate would bear total charges of £40 to £50 per tonne of metal produced at a tin price of £1,500 per tonne, provided there were no deleterious impurities. The smelting charges applying to a 40 per cent concentrate would, on the other hand, amount to over £160 per tonne of metal so produced.

Several other methods, theoretically more efficient, have been devised for the extraction of tin from ores and concentrates. These include reduction-leaching, sulphide or oxide volatilisation (fuming) and chloride metallurgy, any of which might be particularly useful in the treatment of low grade or complex concentrates. While technically attractive, these techniques have yet to acquire general commercial acceptance, although sulphide fuming of low grade concentrates is at present in commercial operation in Bolivia. It is probable that the capital invested in expertise and equipment in conventional processing, as well as the engineering and metallurgical problems involved, have tended to inhibit advances in these fields.

The impact of technological change is likely to apply most radically in areas in which other materials can perform as effectively as, or more effectively than, those containing tin. The problems arising from substitution are broadly dealt with below. The success of tinplate in maintaining and improving its position as a packaging material is notable, and can only be enhanced by the recent development of the drawn and ironed tinplate can which eliminates the necessity for a soldered seam.

*Research*

In order to encourage and expand the use of tin, and to counter substitution, the major tin-producing countries finance the Tin Research Institute, an organisation devoted to promoting new uses for tin and to improving the metal's performance in established uses. In addition to carrying out research work at its laboratories at Greenford, Middlesex, the Institute also sponsors research in universities and other institutions. Recent work has included investigations aimed at optimising the performance of tin plate by further improvement of the inter-metallic bond between steel and tin; the determination of the influence of tin additions in the production of malleable iron castings; a study of the effect of the addition of tin in sintered iron components (this resulted in the discovery, which has aroused considerable commercial interest, that the addition of 2 per cent tin along with 3 per cent copper reduced sintering temperatures from 1,150° to 950°C); and investigations into the properties and uses of aluminium-copper-tin alloys and tin-nickel coatings. Other lines of research are concerned with the improvement of purity of solders for electronic purposes and the use of hydrous stannic oxide and other compounds for the extraction of metals by ion exchange. This could have some application in the recovery of uranium from sea water.

Perhaps the most important lines of investigation, from the point of view of maintaining or increasing future tin consumption, are those relating to the organotins (organic compounds of tin). Organotins tend to suffer in competition with other fungicides, etc by reason of their high cost of manufacture, which is added, of course, to the high price of the metal. A great part of current research is therefore directed at finding the most economic methods for manufacture and establishing optimum formulations on a performance per unit price basis. However, organotins possess the advantage that they break down under natural conditions to inorganic tin compounds which are of proven non-toxicity. The use of organomercurials, on the other hand, is known to set in train a progressive toxicity in the food chain. The dangers arising from the use of mercury may attract legislation to control or ban its use as a fungicide in agriculture and paper-making, thus leaving wide opportunities for organotin usage.

Research on dredging tin from the sea-bed in St Ives Bay has been carried out by Coastal Prospecting Limited (a subsidiary of Union Corporation) and Marine Mining Corporation.

Research in beneficiation is being pursued by the tin mining companies, universities and mining colleges, notably the Camborne School of Mines. A recent development has been that of the CTS Screen, patented by the Cornish Tin Smelting Co Limited and in operation in the United Kingdom and overseas. The company claims that the screens have a high capacity per unit and give good results in splitting at between 20 and 300 mesh.

The Science Research Council and the National Research Development Corporation have financed the development of the Bartles-Mozley Concentrator for recovering fine-grained cassiterite from ores and tailings. The recovery by gravity methods of fine cassiterite in sizes below 50 microns presents an economic problem since the fine material is apt to form an important fraction of crushed ores and a large proportion of the cassiterite content of dumps. Tests with the equipment on tailings hitherto regarded as untreatable due to the presence of cassiterite of sizes less than 20 microns have shown recoveries of 63 per cent and over in the -20 +25 micron range. It is a high capacity concentrator capable of treating over 450 litres of pulp per minute and, at the same time occupies little space as its working surface area of 74.3 sq m is composed of 40 sheets of fibre glass stacked one above the other. The machine has been successfully incorporated in several concentration plants including those of Hydraulic Tin Ltd and Geevor Tin Mines Ltd. The Warren Spring Laboratory is engaged on research into the problems of mineral processing at sea.

There is considerable scope for further work on the occurrence of tin ores in south-west England. Though the broad association of tin deposits with granites is unarguable, the mechanisms whereby they were emplaced at particular locations and specific vertical levels are problematical. The mineralised region has therefore been selected as one of the first in an expanded programme of combined geochemical, geophysical and geological reconnaissance projects being conducted by the Institute of Geological Sciences. This will take into account recent hypotheses which associate occurrences of tin with ancient erosion surfaces and volcanism, and includes, in particular, the search for buried granite cusps by means of detailed gravity surveys.

**Production, consumption and trade**

Production of tin-in-concentrate from mining operations in the United Kingdom amounted to 3,327 tonnes in 1972 compared with 1,816 tonnes in 1971. The rapid rise in production stemmed mainly from the output of the new Wheal Jane mine opened in October, 1971. Until that date the bulk of indigenous primary tin was produced at two mines in Cornwall, namely the South Crofty tin mine owned by St Piran Mines Limited, and Geevor mine, operated by Geevor Mines Limited. Both have been in operation for about 200 years, the latter originally under the name of North Levant, although they have greatly expanded to embrace areas previously mined under other titles. In 1972 South Crofty, the larger of the two, produced 1,572 tonnes of 70 per cent concentrates from 145,000 tonnes of ore, exceeding the previous year's production by 377 tonnes. The output of Geevor Tin Mines Limited for the year ending 31st March, 1973, was 1,110 tonnes of 65 per cent tin concentrates from 111,584 tonnes of ore (including low-grade production of 107 tonnes).

During 1972, its first full year of operation, Wheal Jane Limited, a wholly owned subsidiary of Consolidated Gold Fields Limited, produced 1,260 tonnes of tin-in-concentrate from 133,304 tonnes of ore. Towards the end of 1971 output was initiated by Camborne Mines Limited at Pendarves, on the outskirts of Camborne. In the succeeding year the company produced 430 tonnes of concentrates from 36,941 tonnes of ore treated at its Roscroggan mill (see page 24).

Tin concentrates are also produced by three companies from old mine tailings. Hydraulic Tin Limited, a subsidiary of the Continental Ore Company, operated a plant at Bissoe, near Truro, at which it produced 493.7 tonnes of concentrate yielding 118.2 tonnes of tin in 1972. The sources available to the Hydraulic Tin Company were generally of low grade (less than 1 kg of contained tin per tonne) and presented separation difficulties in terms of mineral assemblage, fine grain or non-liberation of cassiterite: as the more profitable dumps are worked, the remaining material tends to become increasingly intractable and lower in grade. Due to lack of accessible material, therefore, the company stopped production in January, 1973, and put the plant on a care-and-maintenance basis.

Concentrates are also produced from dump material by Tolgus Tin Limited at North Country, near Redruth, with the recovery of 151 tonnes of 32 per cent concentrate during 1972, and by Brea Tin Limited, near Carn Brea village, which produced 98 tonnes of concentrate from 14,735 tonnes of tailings in the same year. The Cornish Tin Smelting Company, which previously recovered concentrates from tailings at its plant at Roscroggan, was taken over by Camborne Mines Limited in March, 1971, and was wholly employed during 1972 in treating ore from Wheal Pendarves on a custom basis. Alluvial cassiterite is produced near Ashburton, east of Dartmoor, by the Glendinning Group of Companies, a local quarrying firm.

With the establishment of Wheal Jane and the programme of expansion currently being carried out at South Crofty, together with the intense exploratory activity undertaken throughout the south-west region in recent years, the prospects for increased tin production are favourable, although the effect of rising costs (bearing in mind that the price of tin is determined on an international basis) cannot be ignored. It is possible that Cornwall may be producing over 5,000 tonnes of tin-in-concentrate per year in the early nineteen-eighties, but even at this level of production domestically mined tin-in-concentrate would meet only about 30 per cent of demand for primary tin.

During the long history of tin mining in south-west England production has fluctuated greatly. Annual production of tin-in-concentrate for the period 1851-1972 is shown in Table 2. The most productive period, the latter half of the 19th century when the area became the world's largest producer, was initiated by the discovery that tin occurred at depth in some lodes beneath previously worked copper, and was made possible by advances in pumping technology which enabled mining to be carried on at these deeper levels. Output started to decline in the last decade of the century due mainly to the discovery and exploitation of more profitable deposits overseas, particularly in the Malay Peninsula. Other reasons included the exhaustion of known ore reserves in some mines and the lack of capital for further development. The company structure of the time encouraged full distribution of profits at frequent intervals so that little was normally retained for mine development, plant modernisation and contingencies. The total quantity of tin-in-concentrate produced in south-west England since mining began probably exceeds 2 million tonnes.

In mid-1973 the tin smelting industry in the United Kingdom, among the largest in the world, was situated at Kirkby, Lancashire, and North Ferriby,

Table 2 United Kingdom: Production of tin-in-concentrate, 1851-1972

| | Tonnes | | Tonnes | | Tonnes |
|---|---|---|---|---|---|
| (Annual average) | | | | | |
| 1851-1860 | 6,503 | 1933 | 1,567 | 1953 | 953 |
| 1861-1870 | 9,554 | 1934 | 2,031 | 1954 | 935 |
| 1871-1880 | 9,809 | 1935 | 2,083 | 1955 | 1,054 |
| 1881-1890 | 9,381 | 1936 | 2,133 | 1956 | 1,083 |
| 1891-1900 | 6,569 | 1937 | 2,018 | 1957 | 1,097 |
| 1901-1910 | 4,657 | 1938 | 2,031 | 1958 | 1,135 |
| 1911-1915 | 5,169 | 1939 | 1,766 | 1959 | 1,208 |
| 1916-1920 | 3,845 | 1940 | 1,645 | 1960 | 1,199 |
| 1921 | 690 | 1941 | 1,532 | 1961 | 1,207 |
| 1922 | 376 | 1942 | 1,386 | 1962 | 1,192 |
| 1923 | 1,037 | 1943 | 1,381 | 1963 | 1,307 |
| 1924 | 2,018 | 1944 | 1,310 | 1964 | 1,274 |
| 1925 | 2,377 | 1945 | 1,009 | 1965 | 1,283 |
| 1926 | 2,364 | 1946 | 863 | 1966 | 1,358 |
| 1927 | 2,635 | 1947 | 833 | 1967 | 1,459 |
| 1928 | 2,805 | 1948 | 876 | 1968 | 1,666 |
| 1929 | 3,323 | 1949 | 912 | 1969 | 1,648 |
| 1930 | 2,528 | 1950 | 903 | 1970 | 1,722 |
| 1931 | 608 | 1951 | 850 | 1971 | 1,816 |
| 1932 | 1,358 | 1952 | 933 | 1972 | 3,327 |

*Sources:* 1851-1946 The Report of the Mineral Development Committee ('Westwood')
1947-1968 Department of Trade and Industry
1969-1972 International Tin Council

Yorkshire. While a progressively larger proportion of total tin production is refined from secondary sources (see Table 3), most of the metal is smelted from imported concentrates largely obtained from Bolivia (see Tables 4 and 5). Tin metal is also imported for blending purposes, mainly from Nigeria, but the United Kingdom remains a net exporter of primary metal. In 1972, exports and re-exports amounted to 13,634 tonnes (see Table 6) valued at over £20 million. A small proportion of this was secondary refined metal some of which, in turn, may have originated in imported wastes.

In 1972, consumption of primary refined tin in the United Kingdom was 14,649 tonnes representing a considerable decline as compared with any one of the previous 25 years (see Table 7). However, even at this reduced level, requirement for metal exceeds current mine capacity by more than four times, and is approximately double domestic mine and secondary output taken together. Tin consuming industries in the United Kingdom are therefore heavily dependent on the continued production of primary refined tin from imported concentrate.

The net annual cost to the balance of payments attributable to trade in tin concentrate, metal and alloys amounted to £10.6 million in 1972 as compared with £17.4 million in 1971 (see Table 8), but it should be noted that much of the tin consumed by industry in the United Kingdom goes to the manufacture of goods and articles for export. The dominance of the tinplate industry as a consumer is illustrated in Table 9 which shows the quantities of tin applied in various end-uses. The importance of the United Kingdom non-ferrous alloy industry as an outlet for tin is notable and contrasts with the estimated pattern of world consumption in which the quantity of tin used for solders exceeds that used in bronze and white alloys.

**Table 3  United Kingdom: Production of primary and secondary refined tin, 1953-72*(a)***

*Tonnes*

| | *Primary* | *Secondary* | *Total* |
|---|---|---|---|
| 1953 | 29,323 | 498 | 29,821 |
| 1954 | 27,916 | 533 | 28,449 |
| 1955 | 27,678 | 476 | 28,154 |
| 1956 | 26,858 | 408 | 27,266 |
| 1957 | 34,722 | 330 | 35,052 |
| 1958 | 31,829 | 1,664 | 33,493 |
| 1959 | 27,041 | 987 | 28,028 |
| 1960 | 26,797 | 1,408 | 28,205 |
| 1961 | 24,841 | 1,934 | 26,775 |
| 1962 | 19,050 | 1,161 | 20,211 |
| 1963 | 17,690 | 1,299 | 18,989 |
| 1964 | 17,119 | 2,506 | 19,625 |
| 1965 | 16,759 | 1,899 | 18,658 |
| 1966 | 17,780 | 1,378 | 19,158 |
| 1967 | 23,691 | 2,606 | 26,297 |
| 1968 | 25,333 | 2,829 | 28,162 |
| 1969 | 26,399 | 2,321 | 28,720 |
| 1970 | 22,035 | 2,466 | 24,501 |
| 1971 | 23,153 | 2,068 | 25,221 |
| 1972 | 21,333 | 4,970 | 26,303 |

*(a)* Owing to a change in the basis of recording, the figures shown separately for primary and secondary production prior to 1958 may not be strictly comparable with those for subsequent years.

*Sources:* International Tin Council
World Bureau of Metal Statistics

**Table 4  United Kingdom: Imports of tin-in-concentrate and primary refined tin, 1963-72**

| | *1* *Tin-in-Concentrate* *Tonnes* | *2* *Primary refined tin* *Tonnes* |
|---|---|---|
| 1963 | 16,465 | 8,053 |
| 1964 | 16,976 | 9,026 |
| 1965 | 15,769 | 9,385 |
| 1966 | 19,671 | 10,109 |
| 1967 | 23,629 | 8,329 |
| 1968 | 23,810 | 9,567 |
| 1969 | 27,010 | 7,157 |
| 1970 | 22,632 | 6,424 |
| 1971 | 18,861 | 7,850 |
| 1972 | 18,122 | 6,280 |

*Sources:* *(1)* International Tin Council
*(2)* HM Customers and Excise

**Table 5**

**(a) United Kingdom: Imports of primary refined tin, by country of origin, 1970-72**

*Tonnes*

| *Country* | *1970* | *1971* | *1972* |
|---|---|---|---|
| Nigeria | 5,863 | 5,598 | 5,128 |
| Netherlands | 10 | 231 | 323 |
| Malaysia | 523 | 1,463 | 613 |
| China | 2 | - | - |
| USA | 26 | 4 | 22 |
| Other | - | 554 | 194 |
| Total | 6,424 | 7,850 | 6,280 |

*Source:* HM Customs and Excise

**(b) United Kingdom: Imports of tin-in-concentrate by country of origin, 1970-72**

*Tonnes*

| *Country* | *1970* | *1971* | *1972* |
|---|---|---|---|
| Bolivia | 19,090 | 14,733 | 13,242 |
| South Africa | 1,126 | 1,326 | 1,302 |
| Argentina | 866 | 656 | 618 |
| Australia | 592 | 1,069 | 1,591 |
| Rwanda | 228 | 170 | 77 |
| Czechoslovakia | 88 | 43 | 19 |
| Tanzania | 80 | 72 | 49 |
| Uganda | 68 | 95 | 78 |
| USA | 59 | 73 | 44 |
| Burundi | 38 | 46 | 30 |
| Peru | 9 | 40 | - |
| France | 3 | 5 | 7 |
| Burma | - | 87 | 27 |
| Other | 385 | 446 | 1,038 |
| Total | 22,632 | 18,861 | 18,122 |

*Source:* International Tin Council

Table 6 United Kingdom: Exports and re-exports of unwrought tin, tin-bearing scrap and waste and tin concentrates, 1963-72

*EXPORTS (a)*

| | *METAL* | | *SCRAP AND WASTE* | |
|---|---|---|---|---|
| | *Quantity* *Tonnes* | *Value* *£'000* | *Quantity* *Tonnes* | *Value* *£'000* |
| 1963 | 8,576 | 7,551 | 80 | 5,631 |
| 1964 | 6,546 | 7,673 | 120 | 8,288 |
| 1965 | 6,789 | 9,339 | 88 | 8,654 |
| 1966 | 6,099 | 7,925 | 99 | 25,827 |
| 1967 | 9,528 | 11,267 | 103 | 62,281 |
| 1968 | 10,686 | 13,274 | 259 *(b)* | 46,082 |
| 1969 | 13,768 | 19,083 | 60 *(b)* | 15,857 |
| 1970 | 12,950 | 19,785 | 32 | 24,741 |
| 1971 | 9,553 | 13,909 | 28 | 19,500 |
| 1972 | 13,634 | 20,233 | 23 | 18,761 |

*RE–EXPORTS*

| | *Quantity* *Tonnes* | *Value* *£'000* | *Quantity* *Tonnes* | *Value* *£'000* |
|---|---|---|---|---|
| 1963 | 222 | 196 | - | - |
| 1964 | 256 | 323 | 132 | 162,692 |
| 1965 | 183 | 265 | 48 | 37,500 |
| 1966 | 90 | 126 | 6 | 6,500 |
| 1967 | 22 | 27 | - | 120 |
| 1968 | 20 | 26 | 17 | 14,510 |
| 1969 | 255 | 362 | 504 | 393,249 |
| 1970 | *n.a.* | *n.a* | *n.a.* | *n.a.* |
| 1971 | *n.a.* | *n.a.* | *n.a.* | *n.a.* |
| 1972 | *n.a.* | *n.a.* | *n.a.* | *n.a.* |

Exports of concentrates in 1971 and 1972 were as follows:

1971 675 tonnes valuedat £226,489

1972 32 tonnes valued at £33,600

*(a)* Includes re-exports from 1970-1972

*(b)* Includes magnesium, except raspings and shacings of uniform size

*n.a.* Information not available

*Source:* HM Customs and Excise

Table 7 United Kingdom: Consumption of primary tin metal, 1947-72

| | *Tonnes* | | *Tonnes* |
|---|---|---|---|
| 1947 | 27,824 | 1960 | 22,140 |
| 1948 | 25,646 | 1961 | 20,567 |
| 1949 | 21,157 | 1962 | 21,783 |
| 1950 | 23,627 | 1963 | 20,967 |
| 1951 | 24,275 | 1964 | 19,561 |
| 1952 | 22,916 | 1965 | 19,565 |
| 1953 | 18,703 | 1966 | 18,721 |
| 1954 | 21,526 | 1967 | 17,634 |
| 1955 | 22,796 | 1968 | 17,420 |
| 1956 | 22,196 | 1969 | 18,059 |
| 1957 | 21,824 | 1970 | 16,950 |
| 1958 | 19,462 | 1971 | 16,425 |
| 1959 | 20,935 | 1972 | 14,649 |

*Source:* World Bureau of Metal Statistics

Table 8 United Kingdom: Balance of trade in tin, 1968-72

| *Imports* | *1968 (£'000)* | *1969 (£'000)* | *1970 (£'000)* | *1971 (£'000)* | *1972 (£'000)* |
|---|---|---|---|---|---|
| Concentrates | 27,687 | 35,219 | 30,425 | 23,297 | 24,150 |
| Tin and tin alloys unwrought | 12,633 | 10,146 | 10,004 | 11,477 | 9,660 |
| Tin and tin alloys wrought | 24 | 45 | 103 | 211 | 111 |
| Scrap and wastes | 585 | 777 | 1,038 | 666 | 1,372 |
| Total | 40,929 | 46,187 | 41,570 | 35,651 | 35,293 |
| *Exports and Re-exports* | | | | | |
| Tin unwrought | 13,300 | 19,445 | 19,785 | 13,909 | 20,233 |
| Tin alloys unwrought | 2,298 | 2,876 | 2,889 | 2,485 | 2,390 |
| Tin and tin alloys wrought | 829 | 855 | 1,543 | 1,234 | 1,395 |
| Waste and scrap | 46 | 16 | 25 | 20 | 19 |
| Concentrates | 15 | 393 | 46 | 226 | 34 |
| Tin oxide | 428 | 503 | 476 | 399 | 584 |
| Total | 16,916 | 24,088 | 24,764 | 18,273 | 24,655 |
| Excess of imports over exports | 24,013 | 22,099 | 16,806 | 17,378 | 10,638 |

*Source:* 1968-72 HM Customs and Excise

**Table 9 United Kingdom: Consumption of tin by end-use, 1968-72**

*Tonnes*

| *Product* | | *1968* | *1969* | *1970* | *1971* | *1972* |
|---|---|---|---|---|---|---|
| Tin plate | | 8,782 | 8,788 | 7,960 | 7,977 | 7,119 |
| | (Copper wire | 487 | 575 | 525 | 457 | 513 |
| Tinning | (Steel wire | 99 | 95 | 93 | 79 | 68 |
| | (Other | 704 | 817 | 804 | 801 | 751 |
| Solder | | 1,365 | 1,484 | 1,533 | 1,456 | 1,633 |
| | (White metal | 3,046 | 2,992 | 2,952 | 2,864 | 2,815 |
| Alloys | (Bronze and gunmetal | 2,220 | 2,337 | 2,218 | 1,916 | 1,808 |
| | (Other | 914 | 871 | 1,125 | 1,111 | 1,367 |
| | (Foil and sheet | 163 | 204 | 75 | 37 | 63 |
| Wrought | (Collapsible tubes | 143 | 135 | 132 | 133 | 133 |
| tin | (Pipes, wire and capsules | 16 | 10 | 11 | 11 | 6 |
| Chemicals and other uses | | 1,517 | 1,341 | 1,190 | 1,246 | 1,575 |
| Total | | 19,456 | 19,649 | 18,618 | 18,088 | 17,851 |
| of which | | | | | | |
| Secondary metal | | 2,036 | 1,590 | 1,668 | 1,663 | 3,202 |

*Source:* World Bureau of Metal Statistics

### *World production*

As the United Kingdom is a leading importer and consumer of tin, a brief account of the world supply situation is necessary in order to set domestic production and consumption in perspective. In 1972 Free World production of tin-in-concentrate was 197,100 tonnes (see Table 10), of which Malaysia, Bolivia, Thailand, Indonesia, Australia, Nigeria and Zaire together produced 178,385 tonnes or over 90 per cent. (Australia is the only major tin-producing nation which is industrialised.) It is notable that mine and primary smelter production is largely concentrated in countries where consumption is low in both absolute and per capita terms. However, Table 11, showing production of primary refined tin by country, illustrates the important position of the United Kingdom which, with Thailand, was in 1970-72 second only to Malaysia as a tin smelter and refiner. Comparison of Tables 10 and 11 clearly demonstrates the importance of Bolivia as a supplier of concentrates: since other major producing countries smelt their own concentrates there is no alternative source capable of supplying concentrates on the scale of present imports into the United Kingdom.

**Table 10: World production of tin-in-concentrate, 1970-72**

| | 1970 | *Tonnes* 1971 | 1972 |
|---|---|---|---|
| Malaysia | 73,794 | 75,445 | 76,830 |
| Bolivia | 30,100 | 30,290 | 32,405 |
| Thailand | 21,779 | 21,689 | 22,072 |
| Indonesia | 19,092 | 19,767 | 21,766 |
| Australia | 8,876 | 9,433 | 12,081 |
| Nigeria | 7,959 | 7,326 | 6,781 |
| Zaire | 6,458 | 6,500 | 6,500 |
| Other | 17,642 | 16,350 | 18,715 |
| Total | 185,700 | 186,800 | 197,100 |

*Source:* International Tin Council

**Table 11 World production of primary refined tin, 1970-72**

| | 1970 | *Tonnes* 1971 | 1972 |
|---|---|---|---|
| Malaysia | 91,495 | 87,095 | 91,001 |
| United Kingdom | 22,035 | 23,153 | 21,333 |
| Thailand | 22,040 | 21,742 | 22,281 |
| Indonesia | 5,190 | 9,218 | 12,010 |
| Nigeria | 8,069 | 7,348 | 6,744 |
| Bolivia | 300 | 6,820 | 6,528 |
| Australia | 5,211 | 6,333 | 7,025 |
| Spain | 3,908 | 4,135 | 4,100 |
| USA | 4,540 | 4,450 | 4,000 |
| Belgium | 4,257 | 3,940 | 3,923 |
| Brazil | 3,100 | 3,424 | 3,583 |
| Zaire | 1,396 | 1,350 | 1,400 |
| Japan | 1,375 | 1,297 | 1,349 |
| Germany (Federal) | 1,195 | 1,169 | 859 |
| Netherlands | 5,937 | 837 | - |
| Other | 3,552 | 3,589 | 3,864 |
| Total | 183,600 | 185,900 | 190,100 |

*Source:* International Tin Council

The continuing supply of tin at prices which will enable the metal to remain a major industrial raw material will depend ultimately on the maintenance of reserves exploitable at reasonable cost. Although attempts to estimate global reserves of any metal are fraught with risk, and published figures are liable to popular misunderstanding, some estimate of availability is generally necessary in order to maintain the confidence of consumers in their sources of supply and to aid producers in planning their long term development, taxation and industrial policy.

The International Tin Council, having as part of its objective the task of reviewing the long term need for development of new tin deposits, set up a Working Party in 1961 to obtain information on various aspects of the industry. The information obtained by the Working Party, mainly by means of a detailed questionnaire issued to producing countries, was published in 1965. That part of the survey devoted to reserves excluded the Communist-dominated parts of the world and also Argentina, Brazil, Burma, Laos, Rhodesia and Rwanda. Within these limitations, it was estimated that reserves amounted to about 4 million long tons metal content at a tin price of £900 per long ton (1965 values). It is worth emphasising that, in excluding Brazil, the estimate takes no account of the very large reserves reputed to occur in the State of Rondonia which have so far been little developed, nor does it include inferred reserves, likely to be extensive, in other parts of the world. A more recent estimate of non-Communist world reserves, published by the United States Bureau of Mines in 1971, is 3,415,000 long tons, although this estimate is not to be directly compared with that of the International Tin Council.

In 1947 the United States set out to establish a stockpile of raw materials which were considered to be vital to its survival in the event of a protracted war and, by 1962, had accumulated *inter alia* a total holding of 353,900 tonnes of tin, of which about 329,000 tonnes had been imported from Malaysia mostly between 1947 and 1956. The need to stockpile tin was regarded as particularly pressing as the United States possesses no workable tin deposits, and the known deposits of the entire North American continent could produce only small quantities of metal. During the nineteen-sixties a significant proportion of the stockpile was sold or otherwise disposed of, but the requirement for strategic purposes (the 'objective') was progressively increased from 188,000 tonnes (185,000 long tons) to 203,000 tonnes (200,000 long tons) in 1963, and to 236,000 tonnes (232,000 long tons) as recently as 1969 when the policy of the United States government was that sufficient material should be available to withstand a 3-year period of siege. In April, 1973, as part of a general modification in policy, the objective was greatly decreased from 236,000 tonnes to 41,000 tonnes (40,500 long tons) leaving a declared surplus of 195,000 tonnes (191,500 long tons). This large quantity of tin, roughly equal to current Free World annual primary production, is likely to be released on the world's markets only over a very long period, with due regard for the interests of producing countries. Nevertheless, the existence of such a surplus cannot be without considerable influence on markets.

### *Statistics*

Details of production of tin concentrates in the UK are given by the Department of Industry in its publication 'Digest of Energy Statistics', but since 1968 the production figures have been combined with those of tungsten (output of which is known to be small). UK overseas trade in tin and tin concentrates is dealt with in the Overseas Trade Statistics of the United Kingdom, published by the Department of Industry.

Similar information on production, consumption and trade is given in 'United Kingdom Mineral Statistics, 1973' prepared by the Institute of Geological Sciences. The International Tin Council publishes the 'Statistical Yearbook' and 'Statistical Supplement' which appear in alternate years and a monthly Statistical Bulletin. These deal comprehensively with production, consumption and trade in tin, tin concentrates and tin products on a world basis. Information relating to the US Stockpile is also given.

'World Metal Statistics', a monthly digest produced by the World Bureau of Metal Statistics, includes a section devoted to production and trade in tin, including secondary refined metal but not of recovery of tin in alloys. The 'Statistical Summary of the Mineral Industry', prepared by the Institute of Geological Sciences, lists figures for production, exports and imports of tin concentrates and metal by country. Tin prices are determined by dealings on the London Metal Exchange which publishes a circular, 'Official Average Monthly Prices'. Prices are quoted daily by leading newspapers.

Statistics of pewter production are regarded as commercially confidential. Full statistics of the production from tin streaming operations are not available through official government channels.

## Trends in demand

The International Tin Council, through its work in statistical and other committees concerned with encouraging tin production and consumption, prepares forecasts of future demand for tin. In a study of prospects for world tin consumption to 1975 published in October, 1971, the Council concluded that consumption would increase by about 0.05 per cent to 1.0 per cent per year on average, but that, in the United Kingdom, it would remain static at about 18,000 tonnes a year.

In 1964, a report by an independent economist, Mr W Robertson, compiled from information collected by a working party of the International Tin Council, gave projections for 1965 and 1970. The report estimated that United Kingdom consumption of primary tin in 1965 and 1970 would be 23,270 and 24,590 tonnes respectively, whereas actual consumption in 1965 was 19,565 tonnes and, in 1970, 16,950 tonnes. The overestimation can be attributed to the fact that the acceleration in the change from hot-dipped to electrolytic coating of tinplate was not foreseen; neither was it anticipated that increasing prices and periodic shortfall would lead consumers to reduce the quantity of tin used in manufacturing processes.

In 1969, Mr Robertson undertook a further study of world market prospects for the Rio Tinto-Zinc Corporation in which he estimated that during the nineteen seventies annual world consumption (excluding the Soviet Bloc) would probably lie in the range 162,570 and 167,650 tonnes. As world consumption outside the Soviet Bloc amounted to 175,197 tonnes in 1968 this projection represents a decline which is attributed to the growing use of tin-free steel in competition with tinplate and to the development of yet thinner coatings on tinplate.

The only longer term projections that appear to be generally available are those included in the US Bureau of Mines estimates of metal consumption in the year 2000: the tentative range for possible world (including Communist) demand is projected to be between 263,000 and 496,000 tonnes as compared with about 251,000 tonnes in 1968.

The factors which govern the future use of tin in preference to other metals and materials are extremely complex and attempts at forecasting even in the short term are hazardous. For example, there must be a limit to the progressive decrease in thickness of the tin coating on tinplate, which has been a major factor in maintaining the viability of tinplate as a competitive packaging material. At present there is no indication as to when this ultimate might be attained or to what extent competing materials might improve in quality or price. The quantity of tin used for the manufacture of tinplate in the United Kingdom continues to decline (see Table 9). The area of tinplate consumed in the period 1965 to 1969 is nevertheless known to have increased by 16 per cent. For most other end-uses except alloys, which in the United Kingdom are second only to tinplate as an outlet, there has been a similar decrease.

In order to ensure as far as is possible that a growing market for tin can be maintained and that dependence on tinplate can be reduced, the aim of research by the Tin Research Institute is to increase the number of end-uses, particularly by finding outlets for tin as an additive in small proportions to other metals. Some diversification has already been achieved. The addition of small amounts of tin (about 0.1 per cent) to cast irons for cylinder blocks, piston rings, clutch plates and die blanks is well established. The use of powdered tin as an aid in the manufacture of sintered iron compacts, a cheap method of making small parts, has been tried with encouraging results. The acceptance of a content of 1 per cent tin in such products would ensure a growing demand for upwards of 1,000 tonnes of tin annually. Additional important uses may grow from recent experimental work on the diffusion of tin into steel. It is reported that steel surfaces so treated show a resistance to rusting which, while not high, is considerably higher than that of untreated steel, and that treated surfaces used for preparation of tinplate showed a superior bond between tin and steel. If the incorporation of tin in steels could be shown to improve their properties, potential demand would be greatly increased, but, to date, studies by the Tin Research Institute indicate that problems of embrittlement have yet to be overcome.

Development of uses for existing and new organotin compounds also offer considerable scope for increasing consumption, particularly in view of the doubts surrounding the environmental effects of organomercurials. In a Code of Practice for the use of plastics in contact with foodstuffs, the British Plastics Federation have approved the use of certain dioctyltin stabilisers up to a maximum of 2 per cent by weight. The acceptance of dioctyltin-stabilised PVC for food containers has led to a great increase in the production of blow-moulded bottles which are likely to become more generally used for mineral water and other beverages.

In these and other ways the position of tin as an important industrial metal is maintained, so far largely by the marketing and improvement of its long-established outlets rather than through the development of new uses, important though these are. Much will depend in future on the ability of the tin-producing countries, most of which are developing countries, to keep supplies in line with demand. Shortages accompanied by high prices would stimulate moves towards substitution which, if successful, would probably be irreversible.

**Substitutes**

A number of materials can be used to substitute for, or even supersede, tin in some end-uses. Periodic shortage of tin, particularly in the USA, and the high price of the metal have combined to encourage efforts at substitution, especially for tinplate. Costs of substitute materials are not directly comparable. For example, aluminium substitutes not for tin, but for tinplate; glass is competitive with both tinplate and tin-bearing PVC; stainless steel has replaced tinned mild steel, not tin, in many applications. Other less tangible factors enter into relative costs such as the fact that tinplate containers are capable of being handled magnetically, while those of aluminium are not. Economic data bearing on future trends in substitution are generally unavailable.

However, the following are examples of ways in which alternative materials compete with, or have replaced, those with a high tin content:

*Tin-free steel.* While little information is available on the extent to which tin-free steel has penetrated the tinplate market, it would seem that in the USA about 10 per cent of steel for packaging is protected by materials other than tin. Tin-free steel is low carbon steel sheet coated with chromic oxide, or treated in a solution of sodium dichromate and phosphate. Of all the factors affecting competition between tinplate and tin-free steel sheet, perhaps that which favours tinplate most effectively is the widespread acceptance of its general reliability. Its brightness in comparison with tin-free steel is also an important property and the new process for producing drawn, seamless cans will further enhance its appeal. At least one US company is reported to be seeking to produce cans from plain steel sheet, protected in use by lacquering only.

*Aluminium.* In the early 1960's the use of aluminium for soft drink canning developed rapidly in the United States where it now accounts for a substantial part of the increasing demand for cans for beer as well as soft drinks. The higher cost of aluminium in the United Kingdom has tended to delay similar developments here, but the projected increases in domestic aluminium production over the next few years may radically alter the situation. In 1968 the United Kingdom used some 53,000 tonnes of aluminium in packaging compared with 465,000 tonnes in the USA, representing a per capita consumption of 0.9 and 2.4 kg respectively. The substitution of aluminium foil for tin foil and the replacement of tinned steel kitchenware by aluminium dates from the early development of markets for aluminium.

*Plastics.* The competition to tinplate offered by plastics has been progressively strengthened due to a mounting price advantage. However, penetration of the tinplate market by PVC products does not necessarily imply a reduction in total tin consumption as many forms of PVC contain up to 1 per cent organotin compounds.

*Cardboard.* Paperboard, cardboard and composites are of only marginal importance as substitutes for tinplate and do not, in themselves, represent a serious threat to tin. Indirectly, however, they have taken over much of the possible growth in tinplate markets due to the popularity of frozen foods which are often sold in containers made partly of cardboard.

*Glass.* Until recently the tendency has been for tinplate to substitute for glass rather than the reverse, as glass is relatively expensive and requires special handling in transport. This trend has recently been partly reversed by the introduction of non-returnable, lightweight glass containers. While the consumer appeal of glass suggests that there may be a limit to substitution by either plastics or metal plate, the outcome of competition between glass and clear tin-bearing PVC is important for the tin industry.

*Stainless Steel.* Stainless steel has displaced tinned steel in kitchenware, and stainless steel tankers have tended to replace tinned containers in the transport of milk.

In numerous other fields tin consumption is at risk or has undergone decline. Copper-lead alloys and tin-aluminium alloys (containing only 20 per cent tin) are being used as bearing metals in modern engines in place of Babbitt metal. Consumption of tin in solder is declining as a whole, although the use of high-tin solders in electronics is expanding. In the motor car industry glass fibre has partly replaced solder for repairing car bodywork and in the long term the introduction of aluminium radiators would obviate the need for soldering joints in the existing types of radiator.

It must be emphasised, however, that the major tin-using industries have been established for a long time and have been subject to threats of substitution almost continually. They have proved their ability to adapt, to improve the competitiveness of their products and to find new outlets for tin.

**Industry**

The tin mining industry in the United Kingdom is represented by the Cornish Chamber of Mines and the Cornish Mining Development Association. The members of the former which are active in Cornwall, excluding English China Clays Limited, are as follows:

*Brea Tin Limited.* This company commenced operations in July, 1971, at its tailings treatment plant at Brea, south of Camborne.

*Camborne Mines Limited.* This was a partnership comprising Strauss Exploration Inc., Anglo Lautaro Nitrate Limited and the Pacific Tin Consolidated Corporation with other, minor holdings. The group brought the new mine, Wheal Pendarves, into production at a proposed rate of about 90,000 tonnes of ore annually towards the end of 1971. Ore was transported for treatment to the Cornish Tin Smelting Company's mill at Roscroggan. Underground development and exploration continued at Wheal Pendarves throughout 1972 but in January, 1973, the company withdrew leaving the operation in the hands of the Official Receiver. In June, 1973, it was announced that St Piran Mines Limited had acquired the mine assets for a sum of £50,000 with the intention of milling Pendarves ore at the South Crofty plant, the capacity of which had been greatly expanded during 1971 and 1972.

*Consolidated Gold Fields Limited.* This company opened Wheal Jane on 1st October, 1971, and the mine is now in full operation. The company also continued exploration in other parts of the south-west region.

*The Cornish Tin Smelting Company Limited.* The company operated a tailings treatment plant until September, 1971, when it was closed down for modification to enable it to process ore from Wheal Pendarves. It formed part of the Wheal Pendarves assets purchased from the Official Receiver by St Piran Mines Limited in May, 1973.

*Cornwall Tin and Mining of New York.* The company, a successor to Prado Exploration Limited (which retains a 45 per cent interest), itself a subsidiary of International Mine Services, completed its feasibility studies on the Mount Wellington property in late 1971. This entailed the sinking of a new shaft about 215 m deep and about 1500 m of lateral exploratory development. The mine was permitted to flood at the beginning of 1972 in order to save costs while financing was completed. Current expectations are that production will begin towards the end of 1974. The company is also undertaking assessment work at the Cligga tungsten-tin deposit, and had an interest in a group (Golden Ram Investments) which, until late 1972, was evaluating the Plantation tungsten-tin lode in East Cornwall.

*Geevor Tin Mines Limited.* This company is at present developing Simm's Lode in the Boscaswell (northern) section of Geevor mine. Further underground development is intended in the abandoned Pendeen Consols property, but it is considered unlikely that the company will seek to increase their output significantly. In December, 1972, the company announced that it had completed evaluation of the old Levant workings following withdrawal of Union Corporation from partnership with Geevor in St Just Mining Services Limited. Results were disappointing and did not justify any attempt at extraction. Development seaward on the Coronation Lode (the Prince of Wales Lode in the Levant Mine), made possible by dewatering Levant to the 400 m level, was started in 1972 with encouraging results.

*Hydraulic Tin Limited.* A subsidiary of the Continental Ore Company Limited, the company was engaged in recovering tin and other minerals from old tailings at its plant near Truro where production was upwards of 100 tonnes of tin-in-concentrate annually. The plant was put on a care and maintenance basis in January, 1973, due to lack of millfeed.

*Penwith Mineral Exploration Limited.* The company have been carrying out extensive exploration on a property near St Just, and are reported to have found a favourable lode. During 1972 the lode was intersected in a crosscut and further examined in new outcrops. The company also reports the discovery of a second lode in an old working on its property.

*South Crofty Limited.* The company, an operating subsidiary of St Piran Mines Limited (formerly Siamese Tin Syndicate), is engaged in a programme of expansion aimed at doubling output from its mine at Pool, near Camborne. The increase in production will come from lodes already developed and the parent company's recent acquisition of Wheal Pendarves will enable some of the millfeed to be drawn from that mine. The company is also pursuing a vigorous programme of reserves extension and is currently driving eastwards into the former East Pool and Agar mine which is being dewatered. Underground drilling is routinely employed to trace extensions of known veins and to locate blind lodes. There is a proposal to drill in the Cooks Kitchen section of the mine in order to seek further extension of major producing bodies at depth.

*Tehidy Minerals Limited.* This company holds mineral rights over more than 14,000 hectares throughout Cornwall and derives most of its revenue in the form of share income and dues from mining rights.

*Tolgus Tin Company.* The company has a tailings treatment plant north of Redruth where low-grade concentrates are extracted from the East

Pool tailings at Tolvaddon, near Camborne, and from beach sand from Upton Towans, Hayle.

*Thyssen (UK) Limited.* The company carried out drilling in the north Godolphin Hill area during 1972 with disappointing results.

Other companies carrying out investigations in south-west England include Marine Mining Corporation (a joint venture by Alpine Geophysical Associates and Bessemer Securities Corporation), Kappa Exploration Limited, Continental Ore Company Limited and Winson Investments Limited. Richardson Mining Associates of Toronto possesses a lease over the tungsten-tin deposit at Hemerdon and were reported to be engaged in clearing old shafts and crosscuts in May, 1972.

A number of small 'tin-streaming' operations, involving recovery of cassiterite from alluvium or tailings, are usually in operation in the south-west region at any given time. Alluvial cassiterite is being recovered in a working near Ashburton owned by the Glendinning Group of Companies Limited, a local quarrying firm. Watts, Blake, Bearne and Company Limited are extracting small quantities of cassiterite from overburden removed during clay mining operations.

The Department of Industry has contact with the mining industry through the Cornish Mining Development Association, the Cornish Chamber of Mines and the main producing companies. Government action to encourage prospecting has included, in particular, the Minerals Exploration and Investment Grant Act, 1972, designed to enable a company to recover from the Exchequer 35 per cent of exploration and evaluation costs on approved projects, repayable if production ensues. Action has also been taken to encourage owners to enter into negotiations with companies seeking to acquire or investigate deposits. This takes the form of tax adjustments permitting mineral rights owners to benefit to a greater extent from royalties.

The main producer of primary refined tin in the United Kingdom, Williams Harvey and Company Limited of Kirkby, Lancashire, was put into liquidation by its parent company, Consolidated Tin Smelters Limited, in June, 1973. It is reported that technical problems with the company's new plant, commissioned only in 1969, resulted in losses which could not be sustained. Traditionally a smelter of high and middle grade concentrates using reverberatory furnaces, the company was a major supplier of tin to United Kingdom consumers, particularly to the British Steel Corporation. Most of the concentrate smelted at the plant is imported from Bolivia which aims to expand its own smelting facilities to process the major part of its concentrate. The only other smelter and refiner, Capper Pass and Son, a subsidiary of the Rio Tinto-Zinc Corporation, processes some low-grade concentrate in addition to slags, sludges and various wastes containing tin and other non-ferrous metals. Processing in the plant at North Ferriby is relatively complex as compared with reverberatory furnace smelting, and is based upon initial roasting and reduction of the metal-bearing charge in blast furnaces, followed by separation and refining of furnace output in several circuits.

Consumers of tin metal and tin-in-alloy are represented by a number of associations including the Tinplate Conference, the British Non-Ferrous Metals Association, the British Tin Box Manufacturers Federation, the

Association of Bronze and Brass Founders, the Tinplate Merchants Association, the Association of British Pewter Craftsmen and several others. Among the many companies manufacturing tin products, the British Steel Corporation dominates the production of tinplate. The three principal can makers are Metal Box Limited, Reads Limited (a subsidiary of the American Can Company) and Crown Cork Limited. Beyond the tinplate and unwrought metal stages, however, outlets for tin diffuse rapidly among the manufacturers of foodstuffs, paints, aerosols, alloys, glass and consumer durables who cannot be regarded as belonging to the tin industry, although they make use of its products.

## References


Annual Report. International Tin Research Council. (Greenford, Middlesex: Tin Research Institute.)

Manual of Tin Mining Companies. (London: Zorn and Leigh-Hunt.) [Annual.]

Mining Annual Review. (London: Mining Journal.)

Statistical Supplement. Tin, Tinplate, Canning. (London: The International Tin Council.) [Annual.]

Statistical Summary of the Mineral Industry. World Production, Exports and Imports. *Mineral Resour. Div., Inst. geol. Sci.* (London: H.M. Stationery Office.) [Annual.]

Tin and its Uses. Quarterly Review of the Tin Research Institute.

Tin International. (London: Tin Publications Ltd.) [Monthly, with statistical supplement.]

Report of the Mineral Development Committee. *Cmd* 7732, *Minist. Fuel Pwr.* 106 pp., figs., refs. (London: H.M. Stationery Office, 1949.) Tin and tungsten, pp. 49-56.

Tin. Its Mining, Production, Technology and Applications. 2nd Ed. C.L. Mantell. 573 pp., figs., photos., refs. (New York: Reinhold Publishing Corporation, 1949.)

Development Plan. County of Cornwall. Report of the Survey. Written Analysis. Part 1 – The County. Part II – The Towns. 290 and 301 pp. resp., figs., maps. (Truro: Cornwall County Council, 1952.)

The metalliferous mining region of south-west England. H.G. Dines (with petrographical notes by J. Phemister.) *Mem. geol. Surv. Engl. Wales,* 1956. 795 pp. (2 vols.), figs., photos., refs., maps.

The mineral potential of Cornwall. K.F.G. Hosking and J.H. Trounson. *Proc. Symp. Future non-ferr. Min. Gt Br. Ire., 1958,* pp. 355-369, fig., refs., map. (London: The Institution of Mining and Metallurgy, 1959.)

Practical considerations in developing old Cornish mines. J.H. Trounson. *Proc. Symp. Future non-ferr. Min. Gt Br. Ire., 1958,* pp.371-382, figs., refs. (London: The Institution of Mining and Metallurgy, 1959.)

Tinplate Handbook. Facts about Tinplate for Buyers and Users. 4th (revised) Ed. W.E. Hoare. 58 pp., figs., photos., refs. (Greenford: Tin Research Institute, 1963.)

Market guide – tin. *E. & M.J. Metal Miner. Mkts,* Dec. 20, 1965, pp. 5-26 photo.

Report on the World Tin Position with Projections for 1965 and 1970. W. Robertson. 160 pp., figs. (London: International Tin Council, 1965.)

The tin mining industry: a growth point for Cornwall. J.C. Goodridge. *Trans. Inst. Br. Geogr.,* 1966, No. 38, pp. 95-104, fig., refs.

Extractive Metallurgy of Tin. P.A. Wright. 228 pp., figs., refs. (Amsterdam-London-New York: Elsevier Publishing Company, 1966.)

A Region with a Future. A Draft Strategy for the South West. South West Economic Planning Council. 154 pp., figs., maps. (London: H.M. Stationery Office, 1967.)

A History of Tin Mining and Smelting in Cornwall. D.B. Barton. 302 pp., photos., refs., maps. (Truro: D. Bradford Barton Ltd., 1967.)

A Technical Conference on Tin, London, 1967. 640 pp. (2 vols.), figs., photos., refs. (London: The International Tin Council, 1968.)

Unroofing of the Dartmoor Granite and possible consequences with respect to mineralisation. J. Dangerfield and J.R. Hawkes. *Proc. Ussher Soc.,* 1969. Vol. 2, Pt. 2, pp. 122-131, refs.

South-west England. 3rd Ed. E.A. Edmonds, M.C. McKeown and M. Williams. *Br. reg. Geol.,* 1969. 130 pp., figs., photos., refs.

Mines and Miners of Cornwall. Parts 1 to 16 inclusive. A.K.H. Jenkin. (Truro: Truro Bookshop, variously 1961 to 1970.)

A Second Technical Conference on Tin, Bangkok, 1969. W. Fox. 1258 pp. (3 vols.), figs., photos., refs. (London: The International Tin Council, 1970.)

Present trends and perspectives in organotin chemistry. G.J.M. van der Kerk. *Chemy Ind.,* 1970, No. 20, pp. 644-647, refs.

The World Tin Market: Prospects for the Late Seventies. W.A. Robertson. 55 pp., figs., refs. ([London]: The Rio Tinto-Zinc Corporation, [1970].)

Patterns of World Tin Consumption, 1957-1968. A. La Spada. 99 pp., figs. (London: The International Tin Council, [1970].)

Pollution and plant evolution. A.D. Bradshaw. *New Scient.,* 1970, Vol. 48, No. 730, pp. 497-500, fig., photos.

Reclamation of toxic metalliferous wastes using tolerant populations of grass. R.A.H. Smith and A.D. Bradshaw. *Nature, Lond.,* 1970, Vol. 227, No. 5256, pp. 376-377, refs.

Exeter Essays in Geography in honour of Arthur Davies. Edited by K.J. Gregory and W.L.D. Ravenhill. 258 pp., figs., photos., refs., maps. (Exeter: University of Exeter, 1971.) The redevelopment of the Cornish tin mining industry: its problems and prospects. J.R. Blunden, pp. 169-184.

Conference on Tin Consumption. 496 pp., figs., photos., refs., (London: The International Tin Council and the Tin Research Institute, 1972.)

United Kingdom Mineral Statistics 1973. R.A. Healing and M.C. Harrison. *Inst. geol. Sci.* 91 pp., figs. (London: H.M. Stationery Office, 1973.)

Tungsten. Compiled by D. Slater. *Mineral Dossier No 5, Miner. Resour. consult Comm.,* 1973. 43 pp., figs., refs.

Produced in England for Her Majesty's Stationery Office by Product Support (Graphics) Limited, Derby
Dd 288255 K12 10/74